Open RAN

Open RAN

Technology and Ecosystem

Vladimir Yanover

WILEY

This edition first published 2025

Registered Offices
John Wiley & Sons, Inc., 111 River Street, Hoboken, NJ 07030, USA
John Wiley & Sons Ltd, New Era House, 8 Oldlands Way, Bognor Regis, West Sussex, PO22 9NQ, UK

For details of our global editorial offices, customer services, and more information about Wiley products visit us at www.wiley.com.

The manufacturer's authorized representative according to the EU General Product Safety Regulation is Wiley-VCH GmbH, Boschstr. 12, 69469 Weinheim, Germany, e-mail: Product_Safety@wiley.com.

Wiley also publishes its books in a variety of electronic formats and by print-on-demand. Some content that appears in standard print versions of this book may not be available in other formats.

Library of Congress Cataloging-in-Publication Data applied for:

Hardback ISBN: 9781394241149

Cover Design: Wiley
Cover Image: © Vasin Lee/Shutterstock

Set in 9.5/12.5pt STIX Two Text by Straive, Pondicherry, India

Contents

Acknowledgments

To my parents, Grigory and Rachel, who were so supportive to my aspirations from my early days. To my wife Lilia standing behind me over long years. To my family: Igor, Uri, Diana, Inbal, Sean, Sophia, Leon, Eitan, and Libby. I love you all.

I'd like to thank my colleagues, with whom I worked in different companies and different standardization bodies. It was a great pleasure working with you.

1

Introduction

CONTENTS

Acronyms and Abbreviations

AI	Artificial Intelligence
AI/ML	Artificial Intelligence/Machine Learning
CA	Carrier Aggregation
CN	Core Network
C-SON	Centralized SON
CU	RAN Central Unit
DC	Dual Connectivity
DL	Downlink
DRB	Data Radio Bearer
D-SON	Distributed SON
DU	RAN Distributed Unit
eNB	LTE Base Station defined in 3GPP
FH	FrontHaul interface
gNB	5G Base Station defined in 3GPP
HO	Handover
H-SON	Hybrid SON
LCM	Life Cycle Management
MDP	Markov Decision Process

Open RAN: Technology and Ecosystem, First Edition. Vladimir Yanover.

ML	Machine Learning
ms	millisecond
Near-RT RIC	Near-Real-Time Radio Intelligent Controller
NF	Network Function
Non-RT	Non-real time
Non-RT RIC	Non-Real-Time Radio Intelligent Controller
NRT	Non-Real Time
NSI	Network Slice Instance
NSSI	Network Slice Subnet Instance
O1	O-RAN management interface (OAM, FCAPS)
O2	O-RAN cloud management interface
O-Cloud	Cloud defined in O-RAN, which, in particular, supports O2 interface
O-CU	O-RAN defined Central Unit
O-CU-CP	O-RAN defined Central Unit – Control Plane
O-CU-UP	O-RAN defined Central Unit – User Plane
O-DU	O-RAN defined Distributed Unit
O-eNB	O-RAN conformant LTE Base Station, which, in particular, supports E2 interface
OFH	Open Fronthaul interface
O-gNB	O-RAN conformant 5G Base Station, which, in particular, supports E2 interface
O-RU	O-RAN defined Radio Unit
PM	Performance Management
PRB	Physical Resource Block in LTE and 5G
RAN	Radio Access Network
RRM	Radio Resources Management
RRU	Remote Radio Unit
RT	Real Time
RU	Radio Unit
SLA	Service Level Agreement
SMO	Service Management and Orchestration framework
SON	Self-Organized Network
SST	Slice/Service Type
TCO	Total Cost of Ownership
TN	Transport Network
UE	User Equipment
UL	Uplink

1.1 What is Open RAN

Technically, Open RAN is a new architecture of the Radio Access Network (RAN), which enables new business models and facilitates technology advances. The new architecture, developed by O-RAN ALLIANCE, can be considered a next phase of the 3GPP 5G RAN architecture with additional split of RAN and advanced RAN programmability.

Comparatively to the 3GPP RAN architecture, O-RAN added a split of the 3GPP Distributed Unit (gNodeB-DU) into O-RAN Distributed Unit (O-DU) and O-RAN Radio Unit (O-RU) and defined Open Fronthaul interface (OFH) between them. Another additional split allows separation of some RAN control functions from the 3GPP Base Station (gNodeB); these control functions are placed in the newly defined controllers: Non-Real-Time RAN Intelligent Controller (Non-RT RIC) and Near-Real-Time RAN Intelligent Controller (Near-RT RIC). Such separation requires that RAN components support a management interface O1 to the Service Management and Orchestration framework (SMO), defined in O-RAN, and E2 interface to Near-RT RIC. Such RAN components are named O-CU, O-DU, etc., according to the corresponding names in 3GPP.

Another aspect of O-RAN architecture innovations is adding integrated support for cloud, which is critical for deployments with cloudified (virtualized) RAN components. This part includes use of O-RAN-compatible RAN accelerators and cloud management.

With new CU/DU/RU split, O-RAN architecture drives further evolution of the business model in the RAN industry. Standardized definition of interfaces between O-RAN components allows purchase of RAN equipment from different suppliers, with e.g. O-CU, O-DU, O-RU, Near-RT RIC coming from different vendors. Such business model allows easier entry of new vendors to the RAN equipment market, especially for the software components, such as O-CU or Non-RT RIC. Entry of new vendors is considered a strong driver for reduction of the RAN Total Cost of Ownership (TCO).

RAN programmability is another strong driver for the TCO reduction. With significant part of RAN intelligence moved to the SMO, Non-RT RIC, and Near-RT RIC, the network operator will be able to modify the behavior of already deployed RAN by making modification in the RAN controllers without touching RAN components O-CU, O-DU, O-RU. The Non-RT RIC supports non-real-time programmability, and the Near-RT RIC allows near-real-time and real-time network control. The Non-RT RIC can control the Near-RT RIC via policy-based A1 interface. Modification of RAN behavior is especially easy and efficient with use of third-party applications: rApps running on the Non-RT RIC platform and xApps running on the Near-RT RIC platform. These applications can be installed and modified after RAN roll-out, which creates additional level of flexibility. RAN programmability allows efficient customization of the network to fit local environment. Besides, it allows permanent network evolution and improvement, enabling flexible service differentiation and creation of new services.

Chapter 1 of this book contains Introduction.

Chapter 2, "Open RAN market drivers," outlines technological and business aspects, which are driving development of the Open RAN ecosystem and corresponding market domain.

Chapter 3, "Open RAN stakeholders," describes the members of Open RAN ecosystem: network operators, vendors, industry bodies. In particular, the chapter provides an overview of O-RAN Alliance structure and activities, such as O-RAN Plugfests. The chapter also contains overview of major Open RAN deployments.

Chapter 4, "Open RAN Ecosystem evolution," outlines evolution of Open RAN technology and provides information on major Open RAN vendors and on most important of commercially available Open RAN solutions.

Chapter 5, "O-RAN Architecture," describes Service Management and Orchestration framework (SMO), Non-Real-Time RAN Intelligent Controller (Non-RT RIC), Near-Real-Time RAN Intelligent Controller (Near-RT RIC), RAN nodes: O-CU-CP, O-CU-UP, O-DU, O-RU, and basic reference points (interfaces), such as A1, O1, O2, E2, and OFH. The chapter also describes the concept of the cloud platform customized for virtualized Open RAN (O-Cloud).

Chapter 6, "Non-real-time RAN programming," addresses the SMO, Non-RT RIC, and rApps. The chapter contains description of A1 interface and examples of protocol operations.

Chapter 7, "Real-Time RAN programming," is focused on Near-RT RIC, xApps, E2 protocols, and services; it contains examples of E2-based optimization procedures.

Chapter 8, "RAN decomposition and cloudification," describes O-DU/O-RU functional split, O-Cloud architecture, O-Cloud management, and hardware acceleration aspects.

Chapter 9, "Network automation in O-RAN," outlines basic concepts of O-RAN network control and optimization and describes how they are applied to several use cases identified by O-RAN specifications.

Chapter 10, "O-RAN Management," contains description of O-RAN management, O1 interface, and Open FH management plane ("M-Plane") interface.

Chapter 11, "AI/ML support in O-RAN architecture," describes O-RAN architecture aspects related to use of Artificial Intelligence/Machine Learning (AI/ML) in RAN automation/optimization, including placement of associated components and the structure of control loops.

1.2 O-RAN ALLIANCE

The key organization in the development of Open RAN technology is O-RAN ALLIANCE™ (www.o-ran.org), an industry forum founded in February 2018 by AT&T, China Mobile, Deutsche Telekom, NTT DOCOMO, and Orange; it was incorporated in Germany in August 2018. O-RAN ALLIANCE has been created by merger of xRAN forum led by AT&T and CRAN forum led by China Mobile.

O-RAN ALLIANCE defines its mission as promotion of Open RAN architecture to drive the RAN ecosystem toward multi-vendor solutions. Besides, transition to Open RAN is supposed to re-shape the RAN industry with more intelligent and virtualized RAN solutions and lower RAN TCO. Specifications developed in O-RAN enable interoperability between RAN components supplied by different vendors [6].

O-RAN ALLIANCE is directly guided by the network operators. The following categories of organizations can participate in the activities of the forum:

- Operator Members
  Mobile operators; all major mobile operators across the globe joined O-RAN ALLIANCE
- Contributors
  This category is mostly for vendor companies
- Academic Contributors
  This category is for educational and scientific organizations

The following are major areas of O-RAN ALLIANCE activities:

- Development of specifications; see information of O-RAN releases in [1–4]
- O-RAN testing, integration, and certification; this area includes O-RAN Plugfests [5]
- Development of open source software implementing Open RAN components; see [7,8]
- Development of White Papers, promoting Open RAN technology

NOTE: The O-RAN specifications are being actively developed. In this book we refer to the Release 3 set of specifications. The latest versions of the O-RAN ALLIANCE specifications are available on the O-RAN ALLIANCE website at https://www.o-ran.org/specifications.

More details on O-RAN ALLIANCE can be found in Chapter 3.

1.3 Open RAN Driving Forces

While O-RAN ALLIANCE is the key organization driving the Open RAN technology; there are some other organizations, which essentially contribute to development of the technology and ecosystem: Telecom Infra Project (TIP, telecominfraproject.com), Open RAN Policy Coalition (www.openranpolicy.org), O-RAN Software Community (docs.o-ran-sc.org). There is also an informal alliance ("MoU Group") created by major European network operators Deutsche Telekom, Orange, Telefonica, Vodafone; later Telecom Italia (TIM) joined the group. These companies published common MEMORANDUM OF UNDERSTANDING ON THE IMPLEMENTATION OF OPEN RAN BASED NETWORKS IN EUROPE [9] and memorandum BUILDING AN OPEN RAN ECOSYSTEM FOR EUROPE [10] in November 2021. They continue work jointly on Open RAN Technical Priority documents, which contain de facto requirements for O-RAN releases.

Open RAN Policy Coalition (www.openranpolicy.org) was created by members of Open RAN ecosystem with mission statement "to promote policies that will advance the adoption of open and interoperable solutions in the Radio Access Network (RAN) as a means to create innovation, spur competition and expand the supply chain for advanced wireless technologies including 5G." Open RAN Policy Coalition consists of about 60 global technology companies. The consortium promotes Open RAN approach internationally, toward regulators and governments.

Multiple governments provide essential support to Open RAN development, in some cases with direct investments; examples: USA, UK, Germany.

More details of these activities can be found in Chapter 3.

References

1 Deutsche Telekom, Orange, Telefonica, TIM, and Vodafone. (2022). Notes on the Open RAN Technical Priorities (Release 2).

2 Deutsche Telekom, Orange, Telefonica, TIM, and Vodafone. (2022). Open RAN Technical Priority "Release 2" Document.

3 Deutsche Telekom, Orange, Telefonica, TIM, and Vodafone. (2023). Notes on the Technical Priorities (Release 3).

4 Deutsche Telekom, Orange, Telefonica, TIM, and Vodafone. (2023). Open RAN Technical Priority Document Release 3.

5 O-RAN ALLIANCE. (2023). Overview of Open Testing and Integration Centre (OTIC) and O-RAN Certification and Badging Program. White Paper.

6 O-RAN ALLIANCE blog. Open and Transparent Way towards Open RAN by the O-RAN ALLIANCE. https://www.o-ran.org/blog/open-and-transparent-way-towards-open-ran-by-the-o-ran-alliance (accessed 9 March 2024).

7 O-RAN Software Community (OSC). Releases. https://wiki.o-ran-sc.org/display/REL (accessed 9 March 2024)

8 Kafka H. (2020). O-RAN Software Community. Bringing 3GPP and Linux Foundation Together. https://o-ran-sc.org/wp-content/uploads/sites/122/2020/12/ORAN_SC_presentation.pdf (accessed 9 March 2024).

9 Deutsche Telekom, Orange, Telefonica, TIM, and Vodafone. (2021). Memorandum of Understanding on the Implementation of Open RAN based networks in Europe.

10 Deutsche Telekom, Orange, Telefonica, TIM, and Vodafone. (2021). Building an Open RAN Ecosystem for Europe to Lead in this Essential Innovation.

2

Open RAN Market Drivers

CONTENTS

Acronyms and Abbreviations

AI	Artificial Intelligence
AI/ML	Artificial Intelligence/Machine Learning
CA	Carrier Aggregation
CN	Core Network
COTS	Commercial-Off-The-Shelf (products)
CU	RAN Central Unit
DC	Dual Connectivity
DL	Downlink
DRB	Data Radio Bearer
D-SON	Distributed SON

Open RAN: Technology and Ecosystem, First Edition. Vladimir Yanover.

DU	RAN Distributed Unit
eNodeB	LTE Base Station defined in 3GPP
gNodeB	5G Base Station defined in 3GPP; also gNB
HO	Handover
HW	Hardware
LCM	Life Cycle Management
ML	Machine Learning
ms	millisecond
Near-RT RIC	Near-Real-Time Radio Intelligent Controller
NF	Network Function
Non-RT	Non-real time
Non-RT RIC	Non-Real-Time Radio Intelligent Controller
NRT	Non-Real Time
NSI	Network Slice Instance
NSSI	Network Slice Subnet Instance
O1	O-RAN management interface (OAM, FCAPS)
O2	O-RAN cloud management interface
O-Cloud	Cloud defined in O-RAN, which, in particular, supports O2 interface
O-CU	O-RAN defined Central Unit
O-CU-CP	O-RAN defined Central Unit – Control Plane
O-CU-UP	O-RAN defined Central Unit – User Plane
O-DU	O-RAN defined Distributed Unit
O-eNB	O-RAN conformant LTE Base Station, which, in particular, supports E2 interface
OFH	Open Fronthaul interface
O-gNB	O-RAN conformant 5G Base Station, which, in particular, supports E2 interface
O-RU	O-RAN defined Radio Unit
PM	Performance Management
RAN	Radio Access Network
RRC	Radio Resource Control
RRM	Radio Resources Management
RRU	Remote Radio Unit
RT	Real Time
RU	Radio Unit
SLA	Service Level Agreement
SMO	Service Management and Orchestration framework
SON	Self-Organized Network
SST	Slice/Service Type

TN Transport Network
UE User Equipment
UL Uplink

2.1 Introduction

This chapter identifies the following market drivers for Open RAN:

- Reduction of Radio Access Network (RAN) Total Cost of Ownership (TCO) leveraged by the following Open RAN innovations:
  - O-RAN architecture, driving expansion of the ecosystem to break vendor lock-in and therefore intensify competition
  - RAN Programmability, significantly extending functionality of already-deployed RAN equipment, which allows post-deployment modification of RAN behavior without new purchase. RAN programmability also enables creation of new services, not foreseen at the time of purchase and deployment.
- Overall RAN performance improvement, based on use of Non-Real-Time RAN Intelligent Controller (Non-RT RIC) and Near-Real-Time RAN Intelligent Controller (Near-RT RIC controller):
  - Wider network visibility
  - Native implementation of non-real-time and real-time control loops
  - Support for hosting of third-party applications: xApps in the Near-RT RIC and rApps in the Non-RT RIC
  - Built-in support for AI/ML functionalities
- Integrated support of RAN cloudification: moving a part of RAN functionality to cloud with reduction of the power consumption, multi-vendor xHaul support.

2.2 Reduction of RAN Total Cost of Ownership

2.2.1 The Role of Total Cost of Ownership

For wireless network operators, Total Cost of Ownership (TCO) of RAN is an important component of the business model, estimated as 60% of the TCO of the whole network. As the RAN TCO constitutes a significant part of overall operator expenses, its reduction is important.

The report [1], jointly published by Wind River and Mason Analysys, predicts that for some deployment models, Open RAN will allow up to 30% RAN TCO reduction. The report identifies the following Open RAN aspects contributing to the TCO reduction:

- Lower cost of Open RAN components (e.g. O-RAN defined Radio Unit [O-RU]), when supplied by competing vendors
- Lower HW cost in case of centralized/virtualized RAN, comparatively to integrated non-cloudified solutions
- Lower cost of network management and optimization, driven by higher level of automation

The report identifies additional benefits of transition to Open RAN, such as supply chain flexibility and extension of the innovation base.

Publications [1–4] provide numerical estimation of potential RAN TCO reduction driven by the following elements of O-RAN technology: entry of new vendors, RAN cloudification, and RAN programmability. The expected Open RAN benefits are addressed below in Sections 2.2.2, 2.2.3, and 2.2.4 correspondingly.

2.2.2 Entry of New Vendors

Legacy networks typically include large areas (sometimes called "markets") where equipment from a single vendor is deployed. Complementary software (e.g. management) often comes from the same vendor. The 3G/4G/5G evolution brought the industry to the structure with small number of RAN suppliers. A RAN supplier usually offers complete RAN solutions, based on fully proprietary hardware and software. The equipment pricing includes profit margins, which many of network operators consider too large.

As different from that, Open RAN stimulates transition to the market, where, instead of integrated RAN solutions, individual RAN components (e.g. O-RUs) are offered. Such transition implies entry of new vendors that offer RAN components rather than integrated solutions. Entry of new vendors facilitates competition and allows deeper specialization of the vendors. Both aspects are believed to drive the profit margins down. Ideally this process (sometimes called "commoditization") will reduce the RAN TCO.

In addition, cloudification of RAN in the O-RAN architecture and introduction of xApps/rApps create a new sector of the market, where RAN software vendors can offer software modules, using APIs standardized in O-RAN.

With wide choice of O-RAN-compliant products, that are individual RAN components, xApps/rApps and other RAN software modules, O-Cloud software, O-Cloud hardware accelerators, network operators can mix and match products, independently chosen in every category. Such freedom of choice can be used by the network operators to bring their RAN to the desired point in the tradeoff between the cost and performance.

One more benefit for network operators, foreseen in mid-term, is additional stability and flexibility of the supply chain, as the operators will be able to mix and match components provided by a wide range of vendors.

2.2.3 RAN Cloudification

Implementation of a legacy 4G/5G RAN node usually is based on highly specialized hardware and software. Transition to Open RAN includes transition of RAN to the cloud defined in O-RAN ("O-Cloud"). Some of O-RAN functions are commonly implemented in software, so their cloudification is comparatively easy: Non-RT RIC, Near-RT RIC, O-CU, and some part of O-DU functionality. The sources [3, 4] estimate that transition of these functionalities to general-purpose hardware potentially can reduce the RAN TCO by 30–45%.

Implementation of lower layers of the O-DU stack (L1 and the lower part of L2) with reasonable price/performance, requires hardware accelerators. O-RAN ALLIANCE defined open APIs for interface between the software and the accelerator hardware. Open APIs enable support of multiple accelerator vendors as well as vendors of software using the accelerators. With O-RAN-compliant accelerator hardware products and software products, O-DU can be implemented on a general-purpose server equipped with accelerator hardware extensions, e.g. with PCIe extension cards. See Chapter 8 for more details.

2.2.4 RAN Programmability

O-RAN architecture enables RAN programmability, based on the Non-RT RIC and Near-RT RIC controllers and on the support of the A1 and E2 interfaces. The A1 interface allows Non-RT RIC controller to modify the RAN's behavior, for example, by setting preferences for use of frequency channels. In particular, an A1 policy may be set in the way that mmWave channels are allocated to user plane data, while low-bandwidth cells are used for control plane (RRC layer) communications. E2 interface enables near-real-time control of the RAN nodes, such as O-CU, O-DU in wide area. Near-real-time and real-time RAN behavior can therefore be controlled by xApps running on the Near-RT RIC platform. As rApps in the Non-RT RIC and xApps in the Near-RT RIC can be developed by the operator internally or purchased from third parties, use of Non-RT RIC and Near-RT RIC allows the operator to modify RAN behavior after purchase and deployment of the RAN. The RAN programmability allows network operator to create and deploy network services, which were not implemented in the purchased RAN equipment. This level of flexibility essentially eliminates single vendor lock and can significantly reduce RAN TCO in mid/long term. Additionally, this level of flexibility creates for network operators a strong potential for faster technology progress in their networks.

More details can be found in Chapter 6, which provides details of non-real-time RAN programmability based on Non-RT RIC. Chapter 7 provides details of near-real-time and real-time RAN programmability based on Near-RT RIC.

2.3 O-RAN Network Optimization Enablers

O-RAN technology enables advanced network optimization solutions, e.g. Energy Efficiency (EE) type of optimization, traffic steering, intelligent QoS support, and more.

O-RAN architecture can support essentially all classic Self-Organized Network (SON) use cases, defined in 3GPP. Under O-RAN architecture, Non-RT RIC and Near-RT RIC have access to any kind of network data that centralized SON (C-SON) and distributed SON (D-SON) have. Also, the RICs are capable of execution of the optimization actions to drive the RAN behavior to desired direction. Near-RT RIC can perform such actions in near-real time.

But O-RAN architecture goes far beyond capabilities of 3GPP SON. First of all, O-RAN architecture combines advantages of both C-SON and D-SON and eliminates their disadvantages. Specifically, SMO/Non-RT RIC enable non-real-time ("slow") optimization control loops via O1 and A1 interfaces, while Near-RT RIC and E2 interface enable near-real-time and real-time ("fast") control loops. O-RAN architecture enables use of combined non-real-time and real-time control loops, linked in the way that non-real-time loop provides guidance to the real-time loop.

Yet another advantage of O-RAN optimization architecture is availability of policy-based control. Non-real-time policies ("A1 policies") can be set to Near-RT RIC to define optimization goals, restrictions, and priorities. In addition, the A1 policies define optimization preferences in terms of resources (e.g., frequency channels), which can be used (or should be used) to achieve the optimization goals. The E2 interface allows setting of real-time policies ("E2 policies") at the RAN nodes ("E2 Nodes"). Such policies are executed by the E2 Nodes in real time, in response to internal RAN events.

O-RAN architecture offers an important advantage in the part of network data collection. There is full support for collection of Performance Management (PM) data defined in 3GPP. The PM data can be collected by the SMO and then made available to the Non-RT RIC, thus can be included into non-real-time control loops. In addition, O-RAN E2 interface specifications enable support for real-time data collection, making this category of network data available in real-time control loops.

In addition to support of control loops and data collection, the O-RAN architecture includes support for AI/ML, which allows significant extension of control and optimization capabilities. The O-RAN architecture defines facilities like ML Training Host, where AI/ML model can be created, trained, and stored. There can be a central training host and/or a local training host(s). Another facility is ML Inference Host where the ML model is used for inference. There are several options for placement of such facilities in O-RAN controllers. Special arrangements are made for transfer of the AI/ML models between the facilities.

More technical details on network automation, control, and optimization in O-RAN architecture can be found in Chapter 9. Information on AI/ML support in O-RAN architecture can be found in Chapter 11.

2.4 Open RAN Challenges

2.4.1 Survey: The Biggest Barriers in Deploying Open RAN

As any technology innovation, development and deployment of O-RAN technology requires investments, efforts and is associated with risks. For example, the White Paper by Deutsche Telekom [5] identifies several challenging aspects: (1) integration between RAN components, supplied by different vendors, (2) regression testing of RAN components, supplied by different vendors and release management, and (3) maintenance of the 2G services in the future.

The survey in the analyst report [6] identified the aspects, which are considered the biggest barriers in deploying Open RAN at scale. Most of the survey respondents (75%) were affiliated with network operators. Table 2.1 shows the survey results, with the corresponding percent of respondents:

Sections 2.4.2, 2.4.3, and 2.4.4 provide some details for the top rows in Table 2.1 and identify measures taken in the Open RAN ecosystems to overcome the challenges. The progress of Open RAN ecosystem shows efficiency of these measures; see, e.g. [7].

Table 2.1 The greatest barriers to deployment of the Open RAN at scale (operators survey)

Security robustness	23%
Vendor interoperability	19%
Cloud native implementation of CU and DU	16%
Readiness of service management and orchestration layer	14%
System integration	14%
Performance of Open Radio Access Network (Open RAN) system	10%
Internal organizational factors	4%

2.4.2 Security in Open RAN

Network operators pay strong attention to Open RAN security. This attention facilitates research, standardization, and testing activities to identify possible threats, perform risk assessment, and improve the Open RAN security.

Open RAN MoU group of operators issued White Paper [8] on Open RAN maturity, security, and energy efficiency; see also [9–13]. In the part of security, the White Paper delivers several messages:

1) Open RAN MoU group of operators cooperates with national authorities, for example, of UK, Germany, and Italy. The cooperation includes sharing information on development of security specifications, testing specifications, and planned deployments.
2) The group collaborates with O-RAN ALLIANCE and GSMA to coordinate O-RAN security testing specifications and provide them as an input to development of European EC NIS standard specifications.
3) Open RAN MoU group commits to deep security testing, based on O-RAN ALLIANCE Release 3 security testing specifications, with strong intention to address future security assurance schemes
4) For Open RAN solutions, the Open RAN MoU group is determined to follow same security assurance and certification processes, which are established for legacy RAN

In their statements, the Open RAN MoU group members refer to European specifications "5G Supplement to the Guideline on Security Measures under the EECC" [14] by the European Union Agency for Cybersecurity (ENISA, www.enisa.europa.eu). ENISA was founded by the European Parliament in 2008. The specifications [14] contain a 5G technology-specific profile complementing the technology-neutral ENISA Guideline on Security Measures under the European Electronic Communications Code (EECC) [15]. The 5G-specific profile recommends technical security measures to be applied by the European countries. The Open RAN MoU group's White Paper [8] identifies the subset of these measures, specifically applicable to O-RAN technology:

- TM03 Ensuring strict access controls
- TM07 Reinforcing software integrity, update and patch management
- TM09 Using EU certification for 5G network components, customer equipment, and/or suppliers' processes
- TM10 Using EU certification for other non 5G-specific ICT products and services (connected devices, cloud services)

In line with these requirements, O-RAN ALLIANCE developed security requirements specifications O-RAN.WG11.Security-Requirements-Specifications-v08.00 [16], with partitioning to specific O-RAN nodes and specific interfaces. For example, regarding the interface E2, this document requires "E2 interface shall support confidentiality, integrity, replay protection and data origin authentication." For common network security tests, O-RAN ALLIANCE supplied O-RAN security test specifications O-RAN.WG11.Security-Test-Specifications-v06.00 [17].

Research report [18] by Rimedo labs provides an overview of the security aspects related to Open RAN.

In May 2023, Quad Critical and Emerging Technology Working Group, funded by the governments of the US, Japan, Australia, and India, published the report [19] on Open RAN Security. The report identifies virtualization as the component of the Open RAN technology with the highest number of possible security threats, with the note that network virtualization is not specific to Open RAN. The report points that only total of 4% of the analyzed security threats are unique to Open RAN, which allows to conclude that introduction of Open RAN does not fundamentally change the security risks, compared to legacy RAN.

The report also states: "Compared to non-disaggregated, non-virtualized RAN, Open RAN has potential security advantages, incl[uding] openly specified, verifiable security controls and capabilities associated to virtualization and cloudification that can help to improve operational security tasks." The report also identifies the mitigation measures and defines the security checklist. The conclusion of the report: "With these mitigation measures, it is possible to ensure equivalent security level compared to traditional RAN."

2.4.3 Vendor Interoperability, System Integration, and Maintenance

On one hand, achieving interoperability between Open RAN components coming from different vendors is more challenging than with classic RAN integrated products. The reason is that there are more interfaces and more vendors participating in the integration process; see, for example [20]. On the other hand, Open RAN technology and ecosystem provide several enablers, which make interoperability easier:

- According to O-RAN standard definitions, Open RAN components must have built-in functionalities, which make integration easier. In particular, such interfaces as E2 contain functions for discovery of the peer entities and functionalities, and for capabilities exchange
- O-RAN conformant components can collect a wide range of network state and performance information, in both non-real time and real time, from the system level, down to the UE level. This information can help in resolving integration issues.
- O-RAN specifications include detail definition of testing procedures

To help integration and reach interoperability, O-RAN ALLIANCE created a net of Open Testing and Integration Centers (OTICs), which are hosted by major Open RAN operators. There are total over 15 OTICs across the globe, including Europe, North America, Japan, and Korea. Along with integration, the OTICs are involved in regular O-RAN PlugFest events, which provide essential help to the participating vendors in the part of interoperability testing. O-RAN PlugFest and certification activities carried by the OTICs are essentially reducing the mass of potential problems, which can show up in the integration process. Details on O-RAN Certification and Badging Program can be found at the O-RAN ALLIANCE website.

DISH Wireless, based on practical experience of deploying Open RAN, identified system integration and security as key challenging areas in the response to the US government agency NTAI [3]. As DISH was its own system integrator, this experience is especially interesting. The company identified the number of integration partners as "dozens." In the DISH network, DUs are deployed either at the RAN sites or in vicinity. The CUs are deployed on the third-party cloud infrastructure, Amazon Web Services (AWS). According to DISH, the biggest challenge is integration and alignment of features roadmap between multiple vendors.

Open RAN ecosystem progress triggered entry of third-party Open RAN integrators, for example, NEC and Fujitsu. Such companies create Open RAN integration labs where vendors, system integrators, and network operators can go over complete cycle of interop testing, from planning to setup of test installations and execution of tests, until full resolution of all identified problems. Similar activities are arranged by major operators, often in cooperation with vendors. For example, Vodafone created Open RAN Test and Integration lab in UK. Deutsche Telekom is leading a consortium of partners jointly funding and using the Open RAN lab established by the German government. The consortium includes operators Telefonica, Vodafone, and vendors Nokia, Capgemini, Rohde & Schwarz.

Along with pre-deployment integration, use of Open RAN network may be complicated in the part of RAN maintenance logistics. In case when significant number of vendors are involved, processing of tickets may be challenging.

2.4.4 Cloud Native Implementation of CU and DU

Cloud native implementation of CU and especially DU is an important condition of RAN virtualization/cloudification, which is an essential part of Open RAN technology. The critical element here is availability of L1 and L2 hardware accelerators for the DU to enable installation of the DU at the general-purpose COTS servers. Such accelerators are already available; many of them in the format of PCIe extension cards carrying dedicated RAN accelerator chipsets. Use of RAN

Smart Network Interface Cards (SmartNICs) enables support of standardized O-RAN interfaces. There is significant number of 5G RAN accelerators and/or chipsets available from such vendors as Qualcomm, Intel, Marvell, and Xilinx. Leading RAN vendors, such as Ericsson and Nokia, include in their portfolio Cloud RAN offering. More details can be found in Chapter 8.

Based on this observation, Open RAN industry can count on availability of cloud native implementation of the RICs, O-CU (O-CU-CP, O-CU-UP), and O-DU.

References

1 Yigit, G., Chappell, C., Bellis, A., and Monniaux, G. (2022). Open RAN Could Deliver up to 30% TCO Savings for Operators with the Right Platform Strategy and Skill Set. Analysys Mason.

2 Azariah, W., Asisi Bimo, F., Lin, C.-W. et al. (2024). A survey on open radio access networks: challenges, research directions, and open source approaches. *Sensors* 24 (3): 1038.

3 DISH Network Corporation. (2023). Comments of Dish Network Corporation Before the Department of Commerce National Telecommunications and Information Administration.

4 GSMA. (2019). 5G-era Mobile Network Cost Evolution.

5 Ledl, P., Lyne, G., Coletti, C. et al. (2023). O-RAN Town: Piloting a High-Power Multivendor Open RAN Solution in a Brownfield Network.

6 Brown, G. (2023). Heavy Reading's 2023 Open RAN Operator Survey. A Heavy Reading White Paper.

7 Smith, P. (2024). 65 Demonstrations of O-RAN Technology and Solutions at MWC23 Showcasing the O-RAN Ecosystem Progress. https://www.o-ran.org/blog/65-demonstrations-of-o-ran-technology-and-solutions-at-mwc23-showcasing-the-o-ran-ecosystem-progress (accessed 9 March 2024).

8 Deutsche Telekom, Orange, Telefonica, TIM, and Vodafone. (2023). Joint MoU White Paper MWC 2023. Open RAN MoU Progress Update on Maturity, Security and Energy Efficiency.

9 Deutsche Telekom, Orange, Telefonica, TIM, and Vodafone. (2021). Building an Open Ran Ecosystem for Europe to Lead in this Essential Innovation.

10 Deutsche Telekom, Orange, Telefonica, TIM, and Vodafone. (2021). Memorandum of Understanding on the Implementation of Open Ran Based Networks in Europe.

11 Deutsche Telekom, Orange, Telefonica, TIM, and Vodafone. (2022). Notes on the Open RAN Technical Priorities (Release 2).

12 Deutsche Telekom, Orange, Telefonica, TIM, and Vodafone. (2023). Notes on the Technical Priorities (Release 3).

13 Deutsche Telekom, Orange, Telefonica, TIM, and Vodafone. (2023). Open RAN Technical Priority Document Release 3.
14 European Union Agency for Cybersecurity (ENISA) (2021). *5G Supplement to the Guideline on Security Measures Under the EECC*, 2e.
15 European Union Agency for Cybersecurity (ENISA) (2021). *Guideline on Security Measures Under the EECC*, 4e.
16 O-RAN.WG11. Security-Requirements-Specifications-v08.00. Security Requirements Specifications.
17 O-RAN.WG11. Security-Test-Specifications-v06.00. O-RAN Security Test Specifications.
18 Rimedo Labs. (2023). The O-RAN Whitepaper 2023 Security in O-RAN.
19 Quad Critical and Emerging Technology Working Group. (2023). Open RAN Security Report.
20 DISH Wireless. (2022). DISH Wireless Systems Integration White Paper. https://www.dishwireless.com/content/dam/pdfs/wi-pdf-whitepaper-systems-integration-enterprise-exploring-thepossibilities.pdf (accessed 9 March 2024).

3

Open RAN Stakeholders

CONTENTS

Open RAN: Technology and Ecosystem, First Edition. Vladimir Yanover.

Acronyms and Abbreviations

AI	Artificial Intelligence
CU	RAN Central Unit
DU	RAN Distributed Unit
eNB	LTE Base Station defined in 3GPP
gNB	5G Base Station defined in 3GPP
HO	Handover
LCM	Life Cycle Management
MIMO	Multiple Input Multiple Output radio technology
ML	Machine Learning
Near-RT RIC	Near-Real-Time Radio Intelligent Controller
Non-RT RIC	Non-Real-Time Radio Intelligent Controller
O1	O-RAN management interface (OAM, FCAPS)
O2	O-RAN cloud management interface
O-Cloud	Cloud defined in O-RAN, that, in particular, supports O2 interface
O-CU	O-RAN defined Central Unit
O-CU-CP	O-RAN defined Central Unit – Control Plane
O-CU-UP	O-RAN defined Central Unit – User Plane
O-DU	O-RAN defined Distributed Unit
O-eNB	O-RAN conformant LTE Base Station, that, in particular, supports E2 interface
OFH	Open Fronthaul interface
O-gNB	O-RAN conformant 5G Base Station, that, in particular, supports E2 interface
RAN	Radio Access Network
RRM	Radio Resources Management
O-RU	O-RAN defined Radio Unit
RAT	Radio Access Technology
RU	Radio Unit
RRU	Remote Radio Unit
SMO	Service Management and Orchestration framework
UE	User Equipment

3.1 Introduction

This chapter contains information of major Open RAN stakeholders: operators, vendors, industry bodies, Standard Development Organizations (SDOs), cooperation between Open RAN businesses, lobbyist alliances, and governmental programs related to Open RAN.

3.2 Open RAN Operators

3.2.1 AT&T

AT&T Inc. is a multinational telecommunications company headquartered in US.

AT&T is known as a founder and leader of the Open RAN community. In February 2018, AT&T, together with CMCC, Deutsche Telekom, NTT DOCOMO, and Orange, jointly announced creation of the O-RAN ALLIANCE with the goal to drive and coordinate world-wide efforts on development and deployment of Open RAN technology. The new industry forum has been formed by merger of the xRAN Forum led by AT&T and the C-RAN ALLIANCE led by CMCC.

Since O-RAN creation, AT&T is playing the leading role in the forum being a member of the Board of Directors and chairing the Board during long time. AT&T is deeply involved in development of Open RAN standards with company representatives chairing several O-RAN Working Groups. AT&T also hosts O-RAN PlugFests, periodic events where Open RAN vendors can do inter-vendor testing, integration, and evaluation of their products.

Before rollout of the AT&T's Open RAN network, the company was involved in technology-oriented trials, for example trial with Nokia where two companies jointly tested Open RAN-compliant RAN Intelligent Controller (RIC) with the native E2 interface [1].

In December 2023 AT&T issued a press release "AT&T to Accelerate Open and Interoperable Radio Access Networks (RAN) in the United States through new collaboration with Ericsson" [2]. In the press release, AT&T announced a 5-year Open RAN plan to achieve the goal of 70% of its wireless network traffic running on Open RAN capable platforms by late 2026. This activity, supported by Ericsson and Fujitsu, aims to establish fully integrated Open RAN sites in 2024. The press release further says: "Beginning in 2025, the company will scale this Open RAN environment throughout its wireless network in coordination with multiple suppliers such as Corning Incorporated, Dell Technologies, Ericsson, Fujitsu, and Intel."

3.2.2 Rakuten Mobile

Rakuten Mobile is a mobile carrier operating in Japan and Europe, offering LTE (4G) and 5G cellular service. It is a Japanese multinational company, wholly owned by Rakuten Group, Inc. Rakuten Mobile was founded in 2018 and launched its cellular service in Japan in 2020. Together with German telecom provider 1&1, Rakuten Mobile launched a mobile network based on Open RAN technology in December 2023; see [3, 4].

Rakuten Mobile's network deployment follows the Open RAN principles:

- Cloudification of the Core and RAN
- Use of open interfaces between RAN components
- Network programmability and automation

The business model of Rakuten Mobile is highly innovative. In particular, the company founded a subsidiary, Rakuten Symphony, Inc, dedicated to development of solutions for telecom industry, such as Open RAN 5G, RIC Platform, Slice Manager, Cloud-Native Orchestrator, Inventory Manager, and Configuration Manager. In addition to use in the company's network, these solutions are offered to network operators to facilitate transition of the industry to Open RAN; see [5]. See Chapter 4 for more information. Rakuten announced creation of an Open RAN test lab in United Kingdom [6].

3.2.3 Vodafone

Vodafone Group is a British multinational telecommunications company with headquarters in United Kingdom. Vodafone owns and operates networks in 21 countries; with partner networks, the company operates in additional 44 countries. Vodafone is an active member of informal "MoU Group" of five European operators actively promoting the Open RAN concept (see Section 3.2.5).

In 2020, Vodafone unveiled its strategic intention for transformation of its mobile access network to Open RAN. In 2021, Vodafone selected several vendors, NEC, Dell, Samsung, Wind River, Capgemini, and Keysight Technologies, with the goal to deliver the first commercial deployment of Open RAN in Europe [21]. In January 2022, Vodafone launched first Open RAN site in the United Kingdom and announced plan of building 2,500 sites in United Kingdom by 2027, with reaching use of Open RAN technology in 30% of its cell sites across Europe by 2030. Later same year, Vodafone signed a cooperation agreement with NTT DOCOMO to harmonize system integration and test processes for Open RAN vendors, so that the vendors would be able to avoid repeating same test procedures for each of the two operators [7].

Following the announced plans, in 2023, Vodafone began mass Open RAN rollout in the United Kingdom, using Samsung's vRAN product supporting 4G and 5G (multi-RAT). Samsung's solution includes RAN software as well as O-RAN-compliant radios and supports Massive MIMO 64T64R. The operator identified it as "Huge network upgrade" [8]. In April 2024, Nokia and Vodafone completed successful Open RAN trial in Italy [9] with Massive MIMO radios.

While going ahead with Open RAN deployment, Vodafone is investing significant efforts into further development of Open RAN technology. In particular,

Vodafone established R&D center in Malaga, Spain, where multiple companies join efforts for development of new 5G and Open RAN technologies [10]. These companies include network operators, such as Verizon, Telefonica, Orange, and T-Mobile, and vendors, such as Nokia, Samsung, Intel, AMD, TSMC, Broadcom, Marvell, and Xilinx.

One of the projects running at the Malaga center is development of Open RAN chipset with Intel. This project is focused on 4th Gen Intel Xeon processors with vRAN Boost. In 2024, it was announced that two companies are working on deployment of the Open RAN equipment based on this processor in the United Kingdom [11].

Vodafone cooperates with Samsung Electronics and Marvell Technology on combining the Samsung's vRAN software with the Marvell's OCTEON Fusion baseband processors for 5G base stations [12]. The target is the industry's first 5 nm chipset for acceleration of processing in the virtualized RAN, particularly in the O-DU.

Vodafone and Qualcomm are involved in joint development of 5G Open RAN infrastructure solutions using the Qualcomm's X100 5G RAN Accelerator Card and Massive MIMO QRU100 5G RAN Platform [13].

3.2.4 1&1 Drillisch

German mobile service provider 1&1 in 2021 announced intention to build a greenfield Open RAN network. Rakuten Communication Platform (RCP) and Altiostar RAN software were selected for the newly established network. Cooperation with Rakuten Mobile was planned for the network build up.

At the beginning of 2023, the operator announced launch of its Open RAN network, in which the Rakuten's platform provides for the access network, core, and cloud capabilities [5]. Rakuten also supplied its orchestration software, which allows automation of the whole network. The network is using cloud infrastructure servers from Dell and Supermicro, routers from Cisco, software from Mavenir, and antennas from CCI.

At the end of 2023, the operator announced start of mobile services [4]; the press release stated that 1&1 is the first network operator in Europe whose network is based entirely on the Open RAN technology.

3.2.5 MoU Group of Operators

"MoU Group" is an informal name of the group of five European operators actively promoting the Open RAN concept: Vodafone, Deutsche Telekom, Telecom Italia, Telefonica, and Orange. The history of the group started in 2021 from joint Memorandum of Understanding (MoU) document [14]; see also the document "Building an Open RAN ecosystem for Europe: to lead in this essential innovation" [15].

The Memorandum sets Open RAN a strategic priority for Europe to establish leadership in 5G and 6G technology development. It refers to the European Commission's 5G Supply Market Trends report (August 2021) that has defined the target of development an open ecosystem with essential progress in development of open 5G specifications.

In view of this, the Memorandum sets the focus on open interoperable network architectures. The Memorandum points to Open RAN as a significant driver of development and innovation in the mobile network industry, with essential advantages for the 5G supply chain. The paper brings attention of European Union policymakers to the following recommendations:

- Political support for Open RAN from the side of EU governmental bodies
- Promote development of an open and interoperable telecommunications products
- Set up a dialogue between the European governments and industry stakeholders leading in support of Open RAN

The Memorandum also called for establishment of a roadmap for network innovation in Europe and creation of a European alliance to facilitate development of next-generation communication technology and infrastructure. To reduce investment risk for EU Open RAN vendors, the MoU Group called the European Commission and national governments for funding for European consortium projects. Special attention was paid to O-RAN standardization, targeting a globally harmonized set of standards for Open RAN.

Subsequently, the MoU Group issued two documents on Open RAN technical priorities for O-RAN Release 2 (March 2022) [16] and Release 3 (April 2023) [17]; see also Release Notes at [18, 19]. These documents were endorsed by O-RAN ALLIANCE as guidance for development of the corresponding specifications.

In 2023, the MoU Group issued the report "Open RAN MoU Progress Update on Maturity, Security and Energy Efficiency" [20], which outlines the status of the Open RAN ecosystem ePl. The report also identifies key areas of focus for the operators, such as advance of Open RAN technology to facilitate efficient deployment in areas with high population density and cooperation with national cybersecurity authorities, such as European Union Agency for Cybersecurity (ENISA). The report called for special attention to energy efficiency of Open RAN components, including O-Cloud infrastructure and O-RU radio circuits.

3.2.6 DISH Wireless

In 2020, DISH Network acquired Sprint's wireless business Boost Mobile to establish it as DISH Wireless, which became the fourth largest wireless carrier in the United States.

Together with Verizon, DISH Wireless is considered a lead US operator in Open RAN deployments. In May 2022 DISH announced planned deployment of 5G Open RAN network, based on Samsung's vRAN, 5G O-RAN-compliant virtualized RAN solution, and O-RAN-compliant Radio Units (O-RUs) across the United States [22]. This made DISH Wireless the US first virtualized, O-RAN 5G broadband network. As different from many other operators, DISH Wireless is building a "Greenfield" Open RAN deployment.

In 2022 DISH Wireless turned on an Open RAN 5G network with 10,000 cell towers, reaching coverage of 20% of the US population. In February 2023, DISH Wireless and Samsung Electronics launched Samsung-supported sites for the DISH 5G network, as the beginning of the operator's nation-wide rollout. Samsung supplied Open RAN-compliant O-RUs along with vRAN, 5G virtualized RAN software [23]. There are also indications of using Fujitsu radios and Mavenir's baseband software. Cooperation with multiple vendors seems to have a big effect on orientation of RAN vendors to Open RAN.

DISH Wireless is active in O-RAN ALLIANCE, which includes hosting of O-RAN PlugFests, where cross-vendor integration and testing are performed.

For promotion of Open RAN technology, DISH Wireless filed opinion to US National Telecommunications and Information Administration (NTIA); the opinion includes call to US to lead the way through targeted investments, with focus on projects that endorse confidence in Open RAN technology. The opinion calls for lowering the entry barriers for Open RAN vendors.

3.2.7 Deutsche Telekom

Deutsche Telekom AG (DT or DTAG) is a German telecommunications company, the largest telecom provider in Europe by revenue. It was established in 1995. Deutsche Telekom was among the O-RAN ALLIANCE founders (2018), and since then it is very active in the Alliance. Deutsche Telekom representatives were chairing the O-RAN ALLIANCE and several O-RAN Work Groups. The company is an active member of the "MoU Group" of operators (see Section 3.2.5), which plays a significant role in coordination of O-RAN technology development and management of O-RAN Releases.

According to [24], in 2022 Deutsche Telekom defined six steps to Open RAN success:

- O-RAN town trial project to gain expertise and drive readiness of Open RAN solutions
- SMO/Automation to avoid any vendor lock-in and manage integration complexity
- Further development of Open RAN ecosystem to create sustainable supplier chain
- Development of skills internally and through O-RAN community

- New operating model to address complexity of Open RAN supply chain, testing, and integration
- Selection of vendors for large-scale Open RAN deployment

At the end of 2023, Deutsche Telekom announced beginning of deployment of Open RAN network with Nokia and Fujitsu in Northern Germany [25]. The Open RAN network cluster is fully integrated into the operator's live commercial network and supports 2G, 4G, and 5G services. Nokia was selected as major RAN vendor with its Open RAN-compliant 5G AirScale baseband solution (O-CU, O-DU) [26]. The Radio Units (O-RUs) are supplied by Fujitsu. Notably, in this deployment O-DU and O-RU are supplied by two different vendors with full support for Open Fronthaul (OFH) interface, which enables interoperation. O-RAN technology will be used for Cloud RAN management, with support for third-party Cloud as a Service (CaaS). The deployment includes Non-RT RIC, Near-RT RIC, SMO.

For Open RAN technology development, Deutsche Telekom established partnership with major Open RAN vendors. For example, in September 2023 Deutsche Telekom completed Open RAN optimization trial in cooperation with Juniper Networks, AirHop, VMware, and Viavi [27]. Non-real-time RAN optimization, based on Non-RT RIC, rApps, and Service Management and Orchestration (SMO) framework, used AI/ML models. The components used in the trial were supplied by multiple vendors. The trial addressed such use cases as Physical Cell Identifier (PCI) optimization and Energy Saving in multi-carrier environment.

3.2.8 Verizon

Together with DISH Wireless, Verizon is considered a lead US operator in Open RAN deployments. The operator started from RAN virtualization [28]. In December 2022, Verizon announced deployment of more than 8,000 vRAN cell sites, with the target of 20,000 [29]. In February 2024, Verizon announced completed deployment of more than 130,000 O-RAN-compliant radios with mMIMO capabilities in previously reported 15,000 O-RAN-compliant virtualized cell sites.

Verizon, together with partners, is progressing toward deployment of Open RAN technology. For example, in February 2024, Verizon, Samsung, and Keysight completed O-RAN conformance and interoperability testing for commercial-grade LTE and 5G NR radios for Verizon's network [30, 31]. In February 2024, Verizon and AT&T announced creation of the ACCoRD consortium [32] for support of Open RAN integration.

3.2.9 NTT DOCOMO

NTT Docomo, Inc. is a Japanese mobile phone operator, a subsidiary of the Nippon Telegraph and Telephone.

NTT DOCOMO co-founded O-RAN ALLIANCE in 2018. In 2021, NTT DOCOMO published a White Paper "5G Open RAN Ecosystem" [33], which outlined Open RAN benefits, challenges, solutions, and other aspects. The operator created its own forum, Open RAN Ecosystem or OREC, which includes Nvidia, Qualcomm, Intel, Mavenir, NEC, Red Hat, Dell, Fujitsu, HPE, VMware, and Xilinx.

In 2022, NTT DOCOMO signed a cooperation agreement with Vodafone to harmonize system integration and test processes for Open RAN vendors so that the vendors would be able to avoid repeating same test procedures for every of the two operators [7].

At the end of 2023, it was announced that Nokia would deploy multi-supplier O-RAN 5G network with NTT DOCOMO [34]. The reported company's goal was to deploy 20,000 5G base stations with O-RAN-compliant Open Fronthaul. The vRAN part of the solution is based on general-purpose Intel Architecture server with NVIDIA RAN accelerator and SDK; the cloud platform has been supplied by Wind River. Fujitsu plays the role of the system integrator and supplies O-RUs with O-RAN-compliant Open Fronthaul interface [35].

3.3 Open RAN Vendors

Open RAN vendors can be classified into several categories.

One category contains "incumbent" vendors, with long standing market share: Ericsson, Nokia, Samsung. Before Open RAN entry (say, before 2020) such vendors usually supplied integrated RAN solutions. The components of such solutions could not be combined with components from other vendors. When priorities of major mobile operators shifted to Open RAN, such vendors started offering Open RAN products. Usually such evolution starts from cloudification (virtualization) of the integrated RAN product and continues to offering individual RAN components, exposing the interfaces, standardized in O-RAN ALLIANCE. This transition ultimately enables use of such components in a multi-vendor RAN.

Another category of RAN vendors includes such companies as NEC, Fujitsu, Capgemini, Radisys, Mavenir with Open RAN offerings that include vRAN software and/or individual RAN components.

Several vendors offer Open RAN hardware accelerators and/or accelerator chipsets: Qualcomm, Marvell, NVIDIA, Intel, AMD(XILINX).

An interesting sub-category consists of network operators, that established in-house development and productization of RAN components, often via their subsidiaries or joint venture with vendors: Rakuten Symphony, Rakuten Altiostar, Vodafone, DOCOMO (OREX) [36].

More information of Open RAN vendors and specific products can be found in Chapter 4.

3.4 O-RAN ALLIANCE

3.4.1 O-RAN ALLIANCE History

O-RAN ALLIANCE (www.o-ran.org) was created in February 2018 by merger of xRAN Foundation led by AT&T and C-RAN Alliance led by CMCC [37]. The following mobile operators are O-RAN founder members: AT&T, CMCC, Deutsche Telekom, NTT DOCOMO, and Orange. Since foundation, O-RAN ALLIANCE grew up to about 300 members, which includes network operators, vendors, universities, and research institutions. According to the O-RAN publication [38], 30 O-RAN operators members provide service to more than 4.5 billion subscribers.

The following categories of organizations can participate in the O-RAN ALLIANCE activities:

- Operator Members
  All major mobile operators across the globe are O-RAN Operator Members
- Contributors
  This category is mostly for vendor companies
- Academic Contributors
  This category is for educational and scientific organizations

O-RAN ALLIANCE has several areas of activity. The focus is on the development of standard specifications for O-RAN architecture, interfaces, management, and other aspects.

Other O-RAN activities include creation and support of Open Testing and Integration Centers, PlugFest activities, Virtual Exhibitions, next-generation Open RAN research.

3.4.2 The Structure of O-RAN ALLIANCE

O-RAN ALLIANCE has the following Work Groups, responsible for development of specifications in the corresponding areas:

- **WG1:** Use Cases and Overall Architecture Work Group

With responsibility for the O-RAN architecture and use cases, WG1 also guides other Work Groups in this area.

- **WG2:** The Non-Real-Time RAN Intelligent Controller and A1 Interface Work Group

This WG is working on specifications, enabling the support of the Non-RT RIC for non-real-time RAN control and optimization and for policy-based control over Near-RT RIC via the A1 interface. The WG2 scope includes also aspects related to the SMO architecture, to rApps and management of AI/ML models.

- **WG3:** The Near-Real-Time RIC and E2 Interface Work Group

The WG3 is working on specifications for Near-RT RIC operations toward RAN components via E2 interface; this includes near-real-time RAN control and optimization and policy-based RAN control.

- **WG4:** The Open Fronthaul Interfaces Work Group

The objective of this WG is to deliver specifications of Open Fronthaul interface, to support multi-vendor DU-RU interoperability.

- **WG5:** The Open F1/W1/E1/X2/Xn Interface Work Group

WG5 is tasked to develop profile specifications for F1/W1/E1/X2/Xn interfaces defined in 3GPP, with the goal of applying the profiles to O-RAN-compliant equipment

- **WG6:** The Cloudification and Orchestration Work Group

The objective of WG6 is to develop specifications for virtualization (cloudification) of O-RAN, define the technology, and provide reference designs for O-RAN network functions to run on commodity hardware platforms, with focus on the O-CU and O-DU.

- **WG7:** The White-box Hardware Work Group

WG6 is tasked with provision of open reference design hardware to be used in O-RAN equipment.

- **WG8:** Stack Reference Design Work Group

The objective of WG8 is development of software architecture, design, and management of software releases for the O-CU and O-DU.

- **WG9:** Open X-haul Transport Work Group

WG9 is tasked to outline the architecture and specify ways of using the transport network: equipment, physical media, and protocols.

- **WG10:** OAM Work Group

The goal of WG10 is development of the OAM requirements, OAM architecture, and specifications for the O1 interface.

- **WG11:** Security Work Group

This WG aims security aspects of the Open RAN network.

In addition to Work Groups, O-RAN ALLIANCE created several Focus Groups and Research Groups:

- **SDFG:** Standard Development Focus Group

SDFG is responsible for contacts to other Standard Development Organizations (SDOs), such as 3GPP

- **IEFG:** Industry Engagement Focus Group

IEFG works on promotion of the O-RAN technology; the purpose is to leverage adoption of the technology across the industry

- **OSFG:** Open Source Focus Group

The scope of OSFG includes establishment and development of the O-RAN Software Community (OSC), definition of strategy, planning of software release

The O-RAN ALLIANCE Board includes representatives of founding members AT&T, CMCC, Deutsche Telekom, NTT DOCOMO, Orange on permanent basis, and elected members: DISH Network, KDDI, Rakuten Mobile, Reliance Jio, Singtel, Telefonica, Telecom Italia, Verizon, Vodafone. The Board approves O-RAN specifications for publication.

Two committees play significant role in management of O-RAN activities:

- Executive Committee (EC)
  The EC assists the Board by identifying development priorities, projects, and releases for the Board to consider and approve.
- Technical Steering Committee (TSC)
  The TSC provides technical guidance to the WGs and reviews O-RAN specifications prior to the Board approval.

3.4.3 O-RAN Releases

To date, O-RAN ALLIANCE completed three releases of the specifications. Historically, the releases were guided by the document "Open RAN Technical Priorities," with issues for Release 2 and Release 3, created by "MoU group" of European mobile operators: [16–19].

While Release 1 was focused on support for the Open Fronthaul (OFH) interface, the second Release was targeting O-Cloud definitions, hardware accelerators, and interoperability over the OFH interface. Release 3 targets full intelligence and automation with the SMO, Non-RT RIC, and Near-RT RIC. This also includes definition of the APIs enabling support of xApps in the Near-RT RIC as a platform.

3.4.4 O-RAN Certification and Badging Program

O-RAN announced launch of Certification and Badging Program in June 2022. The goal of the program is to assure compliance of O-RAN products with corresponding specifications. Such compliance is important for product vendors and network operators. Certification provides the operators with option to reduce breadth and/or depth of pre-deployment testing in case of certified products.

For vendors, O-RAN certificate can be a desired selling point; also, it can facilitate interoperability cooperation between vendors. Certification testing procedures developed by O-RAN ALLIANCE can be used as a template for pre-deployment testing performed by the network operators. O-RAN certificates are awarded to devices defined in O-RAN specifications: Near-RT RIC, O-CU-CP, O-CU-UP, O-DU, and O-RU.

The O-RAN badges are awarded to sets of devices after successful interoperability testing ("IOT Testing") and E2E testing. The O-RAN Open Testing and Integration Centers (OTICs; Section 3.4.5) perform testing and issue O-RAN Certificates or Badges. See "Testing & Integration" page (www.o-ran.org/testing-integration) and the White Paper linked from that page. Test specifications are developed by the corresponding O-RAN Work Groups; for example, WG2 developed the test specifications for A1 interface: O-RAN.WG2.A1TS-v04.00 [39].

There are three main categories of O-RAN certification tests:

- Conformance tests

O-RAN conformance certification tests address specific interfaces of specific components; they are arranged into profiles, such as O-RU Open Fronthaul profile and O-CU O1 profile. Currently the following profiles are supported:

- O-RU Open Fronthaul interface profile
- O-DU (or combined O-DU/O-CU) Open Fronthaul interface profile

- Interoperability (IOT) tests

O-RAN interoperability tests check interoperability between two products, with interfaces between the products specified in O-RAN. There are also tests for interfaces defined in 3GPP that are included into O-RAN IOT testing ("profiled"). Currently the following IOT badges can be awarded:

- O-RU/O-DU pair, connected by Open Fronthaul interface, defined in O-RAN
- O-eNB/O-en-gNB pair, connected by X2 interface, defined in 3GPP and profiled in O-RAN
- O-CU/O-DU pair, connected by F1-C interface, defined in 3GPP and profiled in O-RAN
- A pair of two O-gNBs connected by Xn-C interface, defined in 3GPP and profiled in O-RAN

- End-to-end (E2E) tests

O-RAN E2E tests are applied to a set of O-RAN components to ensure that the set provides sufficient functionality, E2E performance, and security. Currently the following badges can be awarded:

  - O-RU, O-DU, and O-CU included in the E2E O-RAN system

3.4.5 Open Testing and Integration Centers (OTICs)

O-RAN ALLIANCE created a framework for organization of Open Testing and Integration Centers (OTICs) [40]. Usually, an OTIC is operated by an O-RAN ALLIANCE member company. OTIC's activities, for which the OTIC is qualified, may include the following:

- Support of lab and field trials
- Execution of O-RAN compliance, interoperability testing (IOT), and E2E testing
- Awarding O-RAN certificates and badges
- Hosting O-RAN PlugFests (Section 3.4.7)
- Reference setups and testbed
- Promotion of O-RAN technology and development of skills with demos, workshops, tutorials, etc.

The following OTICs have been licensed by the O-RAN ALLIANCE:

- Europe
  - OTIC in Berlin, Germany
  - OTIC in Torino, Italy
  - OTIC in Madrid, Spain
  - OTIC in Paris, France
- Asia and Pacific
  - Asia and Pacific OTIC in Beijing, China
  - OTIC in ritt7layers in Suzhou City, China
  - OTIC and Security lab in Auray Corp., Taiwan
  - OTIC in Singapore
- North America
  - OTIC in Kyrio O-RAN Test and Integration Lab, Louisville, CO, USA
  - OTIC in NYC Metro Area/East (COSMOS)
  - OTIC in the Boston Area (Northeastern University)
  - OTIC in Washington DC/Arlington VA (Virginia Tech)
  - OTIC in Central Iowa (ARA)
  - OTIC in Salt Lake City (POWDER)
  - OTIC at MITRE Corp. in McLean, Virginia, and Bedford, Massachusetts

- Japan
  - Japan OTIC
- Korea OTIC
  - Korea OTIC

3.4.6 O-RAN Software Community

The O-RAN Software Community (OSC, o-ran-sc.org) is a collaboration between Linux Foundation and O-RAN. The OSC goal is open-source implementation of software components of the O-RAN system under the Apache 2.0 License [41]. The OSC Technical Oversight Committee (TOC) is responsible for all technical aspects of the open-source software development, including code storage, tooling, software integration, and testing.

According to [42], the current OSC software release "I" has been released in December 2023. Release "J" is under development, with targeted release date of June 2024. Release "K" is in the stage of requirements definition, targeting completion in December 2024.

The latest releases include the following projects:

- Controllers
  - Service Management and Orchestration Layer (SMO)
  - Non-RT RIC (NONRTRIC)
  - Near-RT RIC X-APPs (RICAPP)
  - Near-RT RIC E2 interface support (RICPLT)
- O-RAN nodes
  - O-CU (OCU)
  - O-DU High (ODUHIGH)
  - O-DU Low (ODULOW)
- General support
  - Management: OAM controller, VES collector (OAM)
  - Infrastructure: O-Cloud reference design (INF)
  - Integration and Test (INT)
  - AIML Framework (AIMLFW)
  - Simulators (SIM)

Taking example of Non-RT RIC, the corresponding project NONRTRIC has already implemented several RIC software components:

- Non-RT-RIC Control Panel
- Information Coordinator Service
- A1 Policy Management Service
- A1 Policy Controller/Adapter
- rApp Manager Service
- RAN Performance Monitoring Functions (file-based PM), etc.

Documentation of the NONRTRIC project also contains description of Non-RT-RIC Use Cases and Non-RT-RIC Test Framework.

3.4.7 PlugFests

O-RAN PlugFest events are organized by O-RAN ALLIANCE with the goal to facilitate progress of the O-RAN ecosystem. At the event, participating vendors perform integration between their products and try to demonstrate interoperation. Other activities include Proof of Concept (PoC) for new solutions and assessment of system performance.

A PlugFest event can go in several geographic locations simultaneously; this is why they are called "Global PlugFests." For example, O-RAN Global PlugFest 2023 used the following venues:

- Hosted by Auray (Taiwan)
- O-RAN Global PlugFest hosted in Europe by Deutsche Telekom, EANTC, EURECOM, Orange, Vodafone in three European labs: EURECOM Lab in Sophia-Antipolis, France, i14y Lab in Berlin, Germany, and Vodafone Central O-RAN Lab in Newbury, United Kingdom
- O-RAN Global PlugFest hosted in North America by:
  - UNH-IOL, Kyrio, COSMOS Lab, and POWDER Lab
  - University of New Hampshire InterOperability Laboratory (UNH-IOL)
  - CableLabs Kyrio OTIC Lab
  - COSMOS Lab and NSF PAWR SITE with the Rutgers University, Columbia University, NYU Wireless
  - POWDER Lab and NSF PAWR SITE with the University of Utah
- O-RAN Global PlugFest hosted by Korea Telecom

Taking example of European venues, there were about 20 participating companies, including T-Mobile, Vodafone, Orange, Nokia, Radisys, Viavi, VMware, Dell, and others. The event used 10 test scenarios, for which devices from different vendors have been integrated and tested. The test scenarios covered:

- E2E integration of devices from different vendors
- xApps integration with Near-RT RICs; xApps functional verification
- rApps integration with Non-RT RICs; rApps functional verification
- Functional and performance testing
- Conformance testing of the Open Fronthaul interface implementation at the O-RU
- O-Cloud platform performance testing

Test scenarios addressed in the North American venues:

- AI-enable Energy Saving rApp and Non-RT RIC

- O2 Interoperability
- DU+CU combo integration and testing
- O2-IMS conformance testing on Red Hat OpenShift O-Cloud
- O-Cloud orchestration
- O-RAN orchestration
- O-RU real-world performance tests
- xHaul testing: E2E connectivity and configuration, throughput and latency measurements, synchronization, priority traffic test

3.5 Telecom Infrastructure Project (TIP)

The Telecom Infrastructure Project (telecominfraproject.com) is a forum of companies and organizations interested in development and deployment of open, disaggregated, and standard-based solutions. Open RAN is one of TIP cooperation areas. By the end of 2023, the Open RAN Project Group completed Release 3.0 of the Technical Requirements and identified several E2E solution blueprints, covering several categories of deployment scenarios, such as Macro Centralized scenario, Multi-Operator Radio Access Network (MORAN) scenario, and others.

The RAN Intelligence and Automation (RIA) subgroup established a project for development of xApps and rApps for implementation of the use cases defined in TIP:

- Traffic steering
- Predictive load balancing
- Power saving by load-adaptive mode
- QoS-based resource optimization
- Inter-cell interference coordination
- SSB Beam set optimization
- Coverage and capacity optimization

The TIP OpenRAN MoU Group includes TIP participant companies that belong to MoU Group (see Sections 3.2.5, 3.4.3). The TIP OpenRAN MoU Group is focused on promotion of the requirements put forward in Open RAN Technical Priority (Release 3).

3.6 ACCoRD Consortium

In February 2024, AT&T and Verizon announced creation of a new consortium for facilitation and coordination of integration efforts across the Open RAN industry. The name of the new forum is Acceleration of Compatibility and Commercialization for Open RAN Deployments Consortium (ACCoRD) [32]. In addition to testing, the scope of the consortium includes performance and security. Among the companies that joined the consortium are major network operators such as NTT DOCOMO and Jio. In the category of vendors, it's Ericsson, Nokia, Samsung, Fujitsu, Mavenir, Dell Technologies, Intel, Radisys, Rakuten, Red Hat, VMWare, and Wind River Systems. The academic members include University of Texas at Dallas, Virginia Tech, Northeastern University, Iowa State University, Rutgers University, and Idaho National Labs.

3.7 Open RAN Policy Coalition

Members of Open RAN ecosystem created Open RAN Policy Coalition (www.openranpolicy.org) with mission statement "to promote policies that will advance the adoption of open and interoperable solutions in the Radio Access Network (RAN) as a means to create innovation, spur competition and expand the supply chain for advanced wireless technologies including 5G." The consortium consists of about 60 global technology companies, including: Airspan, Altiostar, American Tower, Analog Devices, ARM, AT&T, AWS, Benetel, Bharti Airtel, Broadcom, Ciena, Cisco, Dell Technologies, Deutsche Telekom, DISH Network, Facebook, Fujitsu, Google, HP Enterprise, IBM, Intel, JMA Wireless, Juniper Networks, Marvell, Mavenir, Microsoft, NEC, Nokia, NTT, Nvidia, Oracle, Qualcomm, Radisys, Rakuten Mobile, Reliance Jio, Samsung Electronics America, Telefonica, Texas Instruments, U.S. Cellular, Verizon, VMWare, Vodafone, and Xilinx. The consortium promotes Open RAN approach internationally, toward regulators and governments.

3.8 Support from Governments

US and European governments show strong support for transition to Open RAN. In addition to policy statements, they provide direct funding for Open RAN development and testing.

US National Telecommunications and Information Administration (NTIA) announced creation of the Innovation Fund grant program, advancing open wireless network technologies, specifically pointing to open solutions as a target: "The Innovation Fund will promote 5G+ technologies that are secure, open, and virtualized" [43]. The key aspects were defined as follows:

- Facilitate deployment of open and interoperable equipment.
- Provide for integration of multi-vendor networks.
- Identify criteria of open standards compliance.

The first Notice of Funding Opportunity had two focus areas:

- Research and Development (R&D)
- Testing and Evaluation (T&E).

In April 2022, the UK Government published a Policy paper "Open RAN principles" to coordinate steps aiming acceleration of the development of Open RAN solutions [44]. In September 2023, the UK Government announced significant investment in innovative Open RAN connectivity solutions. Dedicated funding is provided to 19 projects targets aiming to prove feasibility and reliability of the Open RAN technology. The funding was supposed to initiate trials of Open RAN solutions in several urban centers and entertainment venues [45].

In November 2021, German Federal Ministry for Transport and Digital Infrastructure (BMVI) announced the projects selected to receive support from the fund, targeting development and testing of Open RAN solutions. One example of the selected projects is Open RAN test lab "i14y" jointly established by Deutsche Telekom, Vodafone, and Telefonica [46]. The lab is supposed to focus on certification of the Open RAN components that are ready for deployment.

References

1 Nokia. (2023). Nokia and AT&T leverage advanced intelligence of the Open RAN compliant near real-time RAN Intelligent Controller with native E2 in successful trial. Press release (20 March).

2 AT&T. (2023). AT&T to Accelerate Open and Interoperable Radio Access Networks (RAN) in the United States through new collaboration with Ericsson. AT&T News (December 04).

3 Le Maistre, R. (2021). German operator 1&1 to build greenfield Open RAN network with Rakuten.

4 Rakuten. (2023). Europe's first Open RAN is fully functional – Rakuten and Mavenir are technology partners in 1&1's cloud-native mobile network. Press release (December 8).

5 Bicheno, S.1&1 claims launch of the first European Open RAN network. https://www.telecoms.com/open-ran/1-1-claims-launch-of-the-first-european-open-ran-network (accessed 9 March 2024).

6 Le Maistre, R. (2022). Rakuten to host Open RAN test lab in UK.

7 NTT DOCOMO. (2022). Vodafone and NTT DOCOMO sign MoU to cooperate to drive greater system integration and service efficiency in Open RAN. Press release (25 October).

8 Vodafone. (2023). Huge network upgrade begins as Vodafone finishes OpenRAN tests. Vodafone News (31 August).

9 Nokia. (2024). Nokia and Vodafone complete successful Open RAN trial in Italy. Press Release (30 April 2024)
10 Vodafone. (2022). Vodafone opens doors to digital future with new European R&D Centre. Press release (31 January).
11 Vodafone. (2024). Vodafone extends collaboration with Intel on research, development and commercialisation of Open RAN. Vodafone News (4 March).
12 Samsung. (2022). Vodafone and Samsung cooperate with Marvell to accelerate Open RAN performance and adoption. Press release (October 25).
13 Qualcomm. (2022). Qualcomm and Vodafone Partnering to Develop and Test Next-Generation 5G Open RAN Infrastructure. Press Release (October 25).
14 Deutsche Telekom, Orange, Telefonica, TIM, and Vodafone. (2021). Memorandum of Understanding on the implementation of Open RAN based networks in Europe.
15 Deutsche Telekom, Orange, Telefonica, TIM, and Vodafone. (2021). Building an Open RAN ecosystem for Europe to lead in this essential innovation.
16 Deutsche Telekom, Orange, Telefonica, TIM, and Vodafone. (2022). Open RAN Technical Priority "Release 2" Document.
17 Deutsche Telekom, Orange, Telefonica, TIM, and Vodafone. (2023). Open RAN Technical Priority Document Release 3.
18 Deutsche Telekom, Orange, Telefonica, TIM, and Vodafone. (2022). Notes on the Open RAN Technical Priorities (Release 2).
19 Deutsche Telekom, Orange, Telefonica, TIM, and Vodafone. (2023). Notes on the Technical Priorities (Release 3).
20 Deutsche Telekom, Orange, Telefonica, TIM, and Vodafone. (2023). Joint MoU White paper MWC 2023. Open RAN MoU progress update on maturity, security and energy efficiency.
21 Paolini, M. (2022). The Open RAN ecosystem gathers steam. The future of the RAN is open. Senza Fili report.
22 Samsung. (2022). DISH Wireless Selects Samsung Electronics for 5G Open Radio Access Network Rollout. Press Release (May 03).
23 DISH. (2023). DISH Wireless Launches Virtual Open RAN 5G Network with Samsung. Press Release (February 22).
24 Le Maistre, R. (2022). Deutsche Telekom's six steps to Open RAN success.
25 Le Maistre, R. (2023). DT kickstarts its Open RAN rollout with Nokia, Fujitsu.
26 Nokia. (2023). Nokia and Deutsche Telekom begin deployment of multi-vendor Open RAN network in Germany. Press release (12 December).
27 Deutsche Telekom. (2023). Deutsche Telekom and partners demonstrate non-real time RAN optimization in a multi-vendor environment. Deutsche Telekom Media (September 1).
28 Verizon. (2023). Virtualization: What it is and how it's shaping Verizon's 5G network. Verizon News (4 May).

29 Verizon. (2022). Verizon expands VRAN leadership position with addition of first ericsson VRAN cell site. Verizon News (8 December).
30 Samsung. (2024). Samsung, Verizon and Keysight Advance Commercial Open RAN With Extensive Conformance and Interoperability Testing. Samsung Newsroom (February 13).
31 Verizon. (2024). Verizon advances O-RAN technology by deploying 130,000 new O-RAN capable radios in its network. Verizon News (5 February)
32 Soni, R. (2024). The Next Step in Bringing Our Open RAN Vision to Life. AT&T Blog (February 12).
33 NTT DOCOMO. (2021). 5G Open RAN Ecosystem Whitepaper.
34 Nokia. (2023). Nokia deploys energy-efficient multi-supplier O-RAN 5G network with NTT DOCOMO. Press release (20 November).
35 Fujitsu. (2023). Fujitsu delivers O-RAN ALLIANCE-compliant 5G virtualized RAN solution for NTT DOCOMO's 5G commercial network services. Press Release (September 27).
36 NTT DOCOMO. (2023). OREX Announces OREX(R) Open RAN Service Lineup. Press release (September 27).
37 AT&T (2018). Industry Leaders Launch ORAN Alliance. Press release (February 27).
38 O-RAN ALLIANCE. Open and Transparent Way Towards Open RAN by the O-RAN ALLIANCE. https://www.o-ran.org/blog/open-and-transparent-way-towards-open-ran-by-the-o-ran-alliance (accessed 9 March 2024).
39 O-RAN.WG2.A1TS-v04.00. O-RAN A1 interface: Test Specification
40 O-RAN ALLIANCE. (2023). Overview of Open Testing and Integration Centre (OTIC) and O-RAN Certification and Badging Program. White Paper.
41 Kafka, H. (2020). O-RAN Software Community. Bringing 3GPP and Linux Foundation Together. https://o-ran-sc.org/wp-content/uploads/sites/122/2020/12/ORAN_SC_presentation.pdf (accessed 9 March 2024).
42 O-RAN Software Community (OSC). Releases. https://wiki.o-ran-sc.org/display/REL (accessed 28 May 2024).
43 National Telecommunications and Information Administration (NTIA). (2023). Innovation Fund Notice of Funding Opportunity #1 (April 12).
44 UK Government. (2022). Policy paper. Open RAN principles.
45 UK Government. (2023). £88 million research and development boost for innovative connectivity to future proof UK mobile network. Press Release (14 September).
46 Kornstadt, P. (2021). Consortium of industry leaders creates new open lab "i14y" to accelerate network disaggregation and Open RAN.

4

Open RAN Ecosystem Evolution

CONTENTS

Acronyms and Abbreviations

AI	Artificial Intelligence
API	Application Programming Interface
CNF	Cloudified Network Function
COTS	Commercial-Off-The-Shelf (products)
CU	RAN Central Unit

Open RAN: Technology and Ecosystem, First Edition. Vladimir Yanover.

DU	RAN Distributed Unit
eNB	LTE Base Station defined in 3GPP
gNB	5G Base Station defined in 3GPP
FDD	Frequency Division Duplex
HO	Handover
LCM	Life Cycle Management
MIMO	Multiple Inputs Multiple Outputs (technology)
ML	Machine Learning
mMIMO	massive MIMO
Near-RT RIC	Near-Real-Time Radio Intelligent Controller
Non-RT RIC	Non-Real-Time Radio Intelligent Controller
O1	O-RAN management interface (OAM, FCAPS)
O2	O-RAN cloud management interface
O-Cloud	Cloud defined in O-RAN, which, in particular, supports O2 interface
O-CU	O-RAN defined Central Unit
O-CU-CP	O-RAN defined Central Unit – Control Plane
O-CU-UP	O-RAN defined Central Unit – User Plane
O-DU	O-RAN defined Distributed Unit
O-eNB	O-RAN conformant LTE Base Station, which, in particular, supports E2 interface
OFH	Open Fronthaul interface
O-gNB	O-RAN conformant 5G Base Station, which, in particular, supports E2 interface
O-RU	O-RAN defined Radio Unit
RAN	Radio Access Network
RRM	Radio Resources Management
QoE	Quality of Experience
QoS	Quality of Service
RACH	Random Access CHannel
RU	Radio Unit
RRU	Remote Radio Unit
SLA	Service Level Agreement
SMO	Service Management and Orchestration framework
SON	Self-Organizing Networks
TDD	Time Division Duplex
TCO	Total Cost of Ownership
UAV	Unmanned Aerial Vehicle
UE	User Equipment

4.1 Introduction

This chapter provides an overview of the Open RAN ecosystem evolution, with focus on the time interval 2021–2024, when the Open RAN technology reached sufficient maturity to enable interoperation between products supplied by different vendors, the Open RAN conformant products appeared on the market, and major carriers began massive Open RAN roll outs.

Technology progress is described in Section 4.2. Paying attention to O-RAN standardization, the section provides overview of the O-RAN Releases, with focus on RAN cloudification, transition to Open Fronthaul, O1, O2, A1, and E2 interfaces exposed by the Service Management and Orchestration (SMO) framework, Non-Real-Time Radio Intelligent Controller (Non-RT RIC), and Near-Real-Time Radio Intelligent Controller (Near-RT RIC). Section 4.2.2 provides an overview of Open RAN product lines offered by major Open RAN vendors. Along with transition of established vendors to Open RAN, technology development triggered appearance of newcomers in the RAN equipment market.

Section 4.3 links Open RAN technology progress to the Open RAN deployments, addressed in Chapter 3.

4.2 Technology

4.2.1 Standardization

4.2.1.1 O-RAN Specifications Releases

Development of O-RAN specifications is organized in terms of releases. The content of releases follows the corresponding versions of the guidance document "Open RAN Technical Priorities," jointly developed by five major European network operators: Deutsche Telekom, Orange, Telefonica, TIM, and Vodafone. This group of operators sometimes is referred to as "MoU group" because of their common Memorandum of Understanding (MoU) on implementation of Open RAN-based networks in Europe [1, 2] published in January 2021. The "Open RAN Technical Priority" documents [4, 6], and the corresponding Release Notes [3, 5] contain requirements to O-RAN Releases 2 and 3.

The following Open RAN releases were defined:

- **First Release R1**: initial interoperability
  Support of Open Fronthaul (OFH) interface was identified as the key R1 feature, with the goal to cover all aspects needed for full interoperability between the O-RU and O-DU. The R1 did not target interoperability between O-CU and O-DU from different vendors via the F1 interface. No cloud support requirements were included.

- **Second Release R2**: interoperability via OFH interface and HW/SW Separation
In addition to full interoperability over the OFH interface, R2 requires support of O-Cloud and interoperability between software and hardware cloud components coming from different vendors. Example of such requirement in [4]: "O-Cloud shall support Acceleration Abstraction Layer (AAL) and provide a set of Common & Open APIs (e.g. inline, look aside) to CNF software for offloading hardware acceleration functions." The Second Release contains requirements for support of cloudification in the O-RAN network functions. For example, R2 specifies containerization support in O-CU and O-DU.
More details on the Second Release can be found in the Open RAN Technical Priority "Release 2" Document [4] and Notes on the Open RAN Technical Priorities (Release 2) [3], published in March 2022.
- **Third Release R3**: full intelligence and automation
In addition to R1 and R2 requirements, the Third Release introduced support of functions and interfaces, enabling network control and optimization by the SMO, Non-RT RIC, and Near-RT RIC. Support of open APIs for xApps in the Near-RT RIC was included. Third Release pays special attention to security and energy efficiency.
More details on the Third Release can be found in Open RAN Technical Priority Document Release 3 [6] and Notes on the Technical Priorities (Release 3) [5] published in April 2023. The summary of the Release 3 can be found in the O-RAN blog record [7]

For every feature, the document "Open RAN Technical Priorities" identifies support level indicators. One of them is the Priority level: P0 = "must have" or P1 = "nice to have." Another one is Priority type, which shows how many of the 5 MoU operators have the corresponding priority: Unanimity, Majority, or Minority.

The White Paper "Open RAN MoU progress update on maturity, security and energy efficiency" published by the MoU Group in 2023 [8] provides a summary of Open RAN technology progress. The document identifies several areas where O-RAN technology is ready for large-scale deployment: single-band O-RUs, non-mMIMO, and low power O-RUs. The technology for mMIMO and multi-band O-RU products are identified as those, which "may require deployment commitments." The White Paper sets the goal of further development of RIC platforms and the corresponding applications ecosystem. In the key area of security, the document provides guidance to apply the mandatory controls defined in O-RAN and 3GPP and follow a "zero trust" approach through the supplier chain. For power saving, the goal set by the MoU Group, was "at least match the energy efficiency of existing mobile sites."

To date, O-RAN specifications address several use cases listed in Table 4.1.

The following Sections 4.2.1.2, 4.2.1.3, and 4.2.1.4 outline the status of most important Open RAN features through O-RAN Releases 1, 2, and 3.

Table 4.1 Use cases addressed in O-RAN specifications

Use case	Description
Context-based dynamic HO management for V2X	An AI/ML-based method to resolve problematic HO scenarios by adjustment of HO sequences at UE level, under the Near-Real-Time Radio Intelligent Controller (Near-RT RIC) control
Flight-path-based dynamic Unmanned Aerial Vehicle (UAV) radio resource allocation	Reducing unnecessary handovers and excessive usage of radio resources, based on UAV flight path information
Radio resource allocation for UAV application scenario	Latency reduction and improvement of radio resource utilization in the O-RAN architecture, under the Non-RT RIC and Near-RT RIC control
QoE optimization	Focused on applications like VR, which are both bandwidth demanding and sensitive to latency. The solution is based on Non-RT RIC and Near-RT RIC joint operations
Traffic steering	Traffic steering is a tool, which allows changing the routes, over which the traffic is delivered with the purpose of achieving better spectrum utilization and/or QoS improvement. The original and alternative routes may differ in selection of cells, frequency channels, Dual Connectivity legs, etc.
Massive MIMO beamforming optimization	In this use case, increased capacity can be achieved using beamforming of the transmitted signals, spatial multiplexing.
RAN sharing	The goal of this use case is to enable joint use of O-RAN infrastructure by multiple network operators. O-RAN specifications enable the participating operators to control the shared resources, using O1, E2, and O2 interfaces.
QoS-based resource optimization	This use case targets support for differentiated QoS, per UE. Differentiation can be based on network slicing
RAN slice SLA assurance	Assurance of network slice SLA in the part, depending on active RAN Network Slice Subnet Instance (NSSI)
Multi-vendor slices	Enabling network slicing in the scenario where O-CU and O-DU are supplied by different vendors
Dynamic spectrum sharing	This use case provides the requirements for support of the 5G Dynamic Spectrum Sharing (DSS) feature in the O-RAN architecture
NSSI resource allocation optimization	Optimization of the resource allocation to the RAN NSSIs under given service requirements, using capabilities provided by the O-RAN architecture
Local indoor positioning in RAN	Support for indoor positioning under the O-RAN architecture

(*Continued*)

Table 4.1 (Continued)

Use case	Description
O-RAN signaling storm protection	Responds to the Signaling Storm challenges using the O-RAN architecture, with support in xApps and/or rApps
Integrated SON function within the O-RAN framework	Support of the classic SON functions in O-RAN architecture: • PCI allocation • Automatic neighbor relations (ANR) • Mobility load balancing (MLB) • Mobility robustness optimization (MRO) • Coverage and Capacity Optimization (CCO) • RACH optimization (RO)
Shared O-RU	Sharing an O-RU between multiple O-DUs operated by same or different carriers
Energy saving	Energy saving is achieved, e.g. by switching off some of carriers / cells, under control of Non-RT RIC and/or Near-RT RIC
MU-MIMO optimization	Capacity improvement by an xApp, which collects information via E2 interface, generates user selection information and precoding coefficients, and sends them to relevant E2 Nodes for use in MIMO operations
Industrial vision SLA assurance	O-RAN-based RAN optimization to fit strict requirements of the industrial vision applications

4.2.1.2 Open Fronthaul

Support of Open Fronthaul (OFH) interface was defined as the key R1 feature, with the goal to cover all aspects needed for full interoperability between the O-RU and O-DU. Already the first version of the architecture document provided fundamental definitions, such as OFH functional split (also known as O-DU/O-RU split 7.2x), OFH data flows, transport and protocol architecture, definitions of the Control Plane ("C-Plane") protocol, User Plane ("U-Plane") protocol, Synchronization Plane (S-Plane) protocol, and definition of the Management Plane ("M-Plane") interface.

In Release 2, the goal was to achieve full interoperability over the OFH interface. Transport requirements for OFH were added.

4.2.1.3 O-Cloud

In Release 2, O-RAN ALLIANCE added a mass of O-Cloud requirements, most significant being:

- Support of Time Synchronization in different virtualization topologies; SyncE support
- O-Cloud Inventory, Monitoring, and Provisioning Services

- O-Cloud Software Management
- O-Cloud Fault Management
- O-Cloud Infrastructure and Deployment Lifecycle Management
- Acceleration modes: Look-aside and Inline; Smart NICs
- Support for widely used software packages, such as Kiali, Grafana, and Prometheus
- Distributed Architecture
- Scale-In/Scale-Out
- Energy consumption KPIs and Power Saving Features
- Secure SMO and O-Cloud Communication
- Local and remote Identification and authentication
- RAN Sharing/Slicing

4.2.1.4 A1 and E2 Interfaces

In Release 2, the following new elements were added:

- rApp security
- A1 Testing
- AI/ML Support
- A1, E2, and Management APIs
- xApp Repository Function
- E2 Testing

In Release 3, the following additional features were introduced:

- E2SM – KPM (REPORT service)
- E2SM – RC (brand new Service Model): REPORT/INSERT/POLICY/CONTROL services
- A1, E2, Management, Enablement APIs

4.2.2 Open RAN Products

4.2.2.1 Introduction

This section provides an overview of Open RAN products offered by leading Open RAN vendors. In most cases, transition of incumbent vendors to Open RAN is based on existing product line; it starts from RAN virtualization (cloudification) and evolves toward OFH and SMO. RAN industry newcomers usually start directly from fully compliant product. For example, if they offer an O-RU, their solution supports OFH.

4.2.2.2 Ericsson

Telefonaktiebolaget LM Ericsson, or simply Ericsson, is a Swedish multinational telecommunications company with headquarter in Stockholm, founded in 1876. In the wireless industry, the company is mostly known as a supplier of cellular

RAN solutions. The company demonstrated technological and market leadership through 3G, 4G, and 5G generations of mobile technology. Ericsson is one of most active contributors to 3GPP RAN groups and to O-RAN ALLIANCE.

In September 2023, Ericsson published a press release "Ericsson takes leadership in industrializing Open RAN" [9]. It was announced that Ericsson would introduce support for Open Fronthaul in the Cloud RAN product line, starting in 2024. The press release said: "Ericsson is driving the industry in all three areas of Open RAN architectures – cloudification, open management with automation, and open fronthaul." In the part of cloudification, Ericsson came up with Cloud RAN, which is a cloud-native software solution. The White Paper "What's next for RAN? Cloud RAN innovation and purpose-built evolution" [10] outlines the company's vision of cloudified RAN and describes co-existence of Cloud RAN with non-cloudified ("purpose-built") RAN.

The part of automation is supported in the Ericsson Intelligent Automation Platform (EIAP), which is an implementation of SMO framework, including Non-RT RIC; see also the White Paper [10]. There is a plan for EIAP to support the A1, O1, O2, and R1 interfaces. Ericsson's product family architecture does not include Near-RT RIC and does not support E2 interface.

At the beginning of 2024, Ericsson issued a press release [11], which confirms Open RAN as one of the company's strategic directions: "A comprehensive, future-proof radio portfolio prepared for Open RAN, offering communications service providers freedom of choice in evolving network architecture."

At the end of 2023, Ericsson was selected by AT&T as the supplier of Open RAN solution [12]. The carrier expects to have fully integrated Open RAN sites operating in coordination with Ericsson and Fujitsu, starting in 2024. The AT&T's goal is to have 70% of its wireless network traffic to flow across Open RAN capable platforms by the end of 2026.

In February 2024, Ericsson announced that O2 Telefonica in Germany will start deploying Ericsson's Cloud RAN in the first half of 2024, which is considered the deployment of the Cloud RAN product in Europe [13].

4.2.2.3 Nokia

Nokia Networks, formerly Nokia Solutions and Networks (NSN) and Nokia Siemens Networks (NSN), is a multinational telecommunications equipment company headquartered in Espoo, Finland. Nokia is an active member of O-RAN ALLIANCE, with significant contributions to the standard specifications in 3GPP and O-RAN.

In February 2023, Nokia announced launch of an Open RAN product suite called "anyRAN" [14], which includes AirScale cloud-native 5G baseband solution. Due to support of Open RAN architecture, the baseband can be integrated with third-party RAN components, e.g. O-RUs. The suite also includes AirScale

Remote Radio Head (RRH), which can be integrated either with AirScale baseband or with third-party Open RAN baseband solutions. Other components of the suite are AirScale Massive MIMO Adaptive Antenna and NetAct Open, a cloud-native network management system that provides for end-to-end network orchestration and automation. The NetAct Open can be used to manage third-party O-RAN components when needed. The vendor claims that AnyRAN allows gradual evolution from purpose-built RAN to fully cloudified open RAN, with having both at interim steps.

At the end of 2023, Deutsche Telekom announced beginning of deployment of Open RAN network in Germany, where Nokia was selected as major RAN vendor with its Open RAN-compliant 5G AirScale baseband solution (O-CU, O-DU) [15]. The Radio Units (O-RUs) are supplied by Fujitsu. As O-DU and O-RU components are supplied by two different vendors, it can be concluded that these components have fully interoperable implementation of Open Fronthaul (OFH).

The White Paper "Nokia Cloud RAN and Open RAN Solution Roadmap" [16] outlines key elements of the Nokia's Cloud RAN solution. Nokia is using so called "In-Line" acceleration technique assuming it is better suited to provide performance in line with purpose-built RAN, but with improved energy efficiency, and ultimately, lower Total Cost of Ownership (TCO); see also [17].

In [18], Nokia outlined its Near-RT RIC solution and listed AI/ML-enabled xApps to be supplied: Nokia Advanced Traffic Steering xApp and Nokia Anomaly Detection xApp.

4.2.2.4 Samsung

Samsung Electronics Co., Ltd. is a South Korean multinational corporation headquartered in Suwon, South Korea. The company is an active member of O-RAN ALLIANCE. Samsung started Open RAN journey by development of virtualized 5G RAN product, called "vRAN" (https://www.samsung.com/global/business/networks/products/radio-access/virtualized-ran/), with the third phase of the vRAN announced in February 2023 [19]. The solution includes virtualized RAN components vCU, vDU, and Radio Unit (RU), with full support of O-RAN interfaces. The solution can be integrated with mMIMO-capable Radio Units.

In June 2021, Samsung and Vodafone UK announced that Vodafone selected Samsung's vRAN for a wide-scale 5G Open RAN deployment, considered the first such deployment in Europe. Actual deployment of vRAN in Vodafone's network started in 2022 from the area, located in Bath, with more than 2,500 sites. In August 2023, Vodafone and Samsung began mass Open RAN rollout across the United Kingdom [20]. After successful UK launch, Samsung and Vodafone extended deployment of 5G Open RAN networks to Germany and Spain [21].

In February 2023, it was announced that DISH Wireless would launch the US nation-wide rollout of 5G Open RAN Network with Samsung's vRAN [22].

In February 2024, Verizon and Samsung completed O-RAN conformance and interoperability testing for commercial-grade 5G radios for Verizon's network [23].

In February 2024, TELUS issued a press release on partnership with Samsung aiming deployment of the Canada's first 5G Open RAN network [24].

4.2.2.5 Rakuten

Rakuten Symphony, Inc. (symphony.rakuten.com) is a subsidiary of Rakuten Mobile, a mobile operator operating in Japan and Europe. The subsidiary targets development of solutions for telecom industry, which resulted in such products as Open RAN 5G, RIC Platform, Slice Manager, Cloud-Native Orchestrator, Inventory Manager, and Configuration Manager. In February 2023, Rakuten Symphony announced its next-generation Distributed Unit (DU) platform for Open RAN. With Intel's Xeon processors and Intel's vRAN Boost L1 accelerator, the platform allows hosting of computationally challenging DU functions.

Another subsidiary of Rakuten, Altiostar, acquired in 2021, is headquartered in United States. The Altiostar product is Virtual RAN (Cloud RAN) software implementing 4G and 5G.

Products of both Symphony and Altiostar are used in Rakuten Open RAN deployments. Also, they are offered to other network operators to facilitate transition of the industry to Open RAN. German operator 1&1 announced use of Rakuten Open RAN products in their greenfield deployment; see [25, 26].

Rakuten Symphony and its parent, Rakuten Mobile, established an Open RAN Customer Experience Centre in the United Kingdom aiming to promote disaggregated, multivendor radio access network architectures as a viable option for operators and technology developers across Europe and the Middle East [27].

Rakuten is active in O-RAN ALLIANCE, in particular, hosting O-RAN PlugFests.

4.2.2.6 NEC

Nippon Electric Company (NEC) or NEC Corporation is a Japanese multinational information technology and electronics corporation, headquartered at Tokyo, Japan, established in 1898.

NEC Network Products catalog includes two categories of Open RAN products: Open Radio Units (O-RUs) and Telco Cloud [28]. The NEC O-RUs portfolio includes the following sub-categories: high-performance Massive MIMO Radios, FDD Low, and Mid-band Radios, TDD Mid-band Radios. The Telco Cloud product is an O-Cloud platform, optimized for support of vRAN, which includes a COTS hardware, RAN accelerators hardware, and virtualization software, compliant with O-RAN specifications.

NEC Open RAN products are used in commercial deployments by Vodafone, NTT DOCOMO (see Chapter 3).

NEC is active in standards development in 3GPP and O-RAN ALLIANCE; the company is also a member of Open RAN Policy Coalition (Chapter 3).

4.2.2.7 Fujitsu

Fujitsu Limited is a Japanese multinational corporation, established in 1935 and headquartered in Tokyo. The company offers complete Open RAN solution, which includes O-RAN-compliant 5G Radio Units (O-RUs), virtualized O-CUs and O-DUs, and SMO framework implementation. Fujitsu also offers multivendor O-RAN integration services, including lab deployment and conformance testing. Fujitsu participates in Open RAN trials and deployments with multiple operators, such as AT&T [12], Deutsche Telekom [15], NTT DOCOMO [29].

Fujitsu is active in O-RAN ALLIANCE; in particular, the company participates as a vendor in O-RAN PlugFests.

4.2.2.8 Open RAN Software Vendors

Progress of Open RAN ecosystem triggered appearance of RAN software vendors.

- Capgemini
  Capgemini Engineering (previously known as Altran Technologies) is a global innovation and engineering consulting firm headquartered in France. Capgemini has developed RATIO O-RAN Intelligent Controller (RIC), which contains Non-RT RIC and Near-RT RIC, integrated with the Capgemini's AI/ML platform. The xApps and rApps running on the platform support such RAN optimization use cases as slice assurance and energy saving. In June 2021, Vodafone identified Capgemini as one of the key vendors for planned Open RAN deployment in United Kingdom [30]. In December 2022, Vodafone started actual rollout, considered the first Open RAN deployment in Europe [31].
- Mavenir
  DISH is using Mavenir's baseband software solution for deployment of a greenfield Open RAN network in United States [32]. Mavenir supplies a software implementation of virtualized Open RAN Centralized Unit (O-CU) and Distributed Unit (O-DU); the software is capable of operating on vendor-neutral COTS servers.
- Radisys
  Radisys offers O-RAN-compliant software solution (https://www.radisys.com/solutions/openran), which can be used in both bare metal and containerized environment. The solution contains virtualized Centralized Unit (vCU), Distributed Unit (vDU), remote Radio Unit (O-RU), RIC, and centralized management and orchestration solution.
- Intel
  Intel FlexRAN is an Open RAN software RAN platform developed by Intel (https://www.intel.com/content/www/us/en/developer/topic-technology/edge-5g/tools/flexran.html).

- Microsoft
  Microsoft developed RAN Analytics and Control solutions for virtualized RAN running on Microsoft Edge platforms, as a part of Azure for Operators product [33]. The solution includes AI/ML components running on the Non-RT RIC platform (rApps) and Near-RT RIC platform (xApps). For tight-real-time control (1–10 ms), the solution provides inference pieces of code ("codelets") that can be deployed at various points in the RAN software stack. The codelets are executed inline in the live RAN system. They enable network data collection and execution of real-time inference and control decisions. Target areas include interference detection and mitigation, RAN-agnostic dynamic power saving, and more. Microsoft's cooperation with Capgemini and Intel targeted integration of the Microsoft's solution with Capgemini's 5G RAN and Intel's FlexRAN reference software.

4.2.2.9 Open RAN Chipset and Accelerator Vendors

Open RAN ecosystem progress is strongly driven by Open RAN chipset and RAN accelerators vendors. The accelerators are deployed in O-Cloud to provide for reasonable balance between RAN performance and TCO.

- Qualcomm
  Qualcomm's X100 5G RAN Accelerator Card is based on the Qualcomm's chipset. It is an O-RAN-compliant PCIe inline accelerator card, supporting sub-6 GHz and mmWave baseband solution [34]. The solution supports 5G L1 (mMIMO 64T64R) and contains integrated O-RAN fronthaul interface. Qualcomm's QRU100 5G RAN Platform contains 5G baseband, transceiver, and mmWave front-end with antenna module. It is an O-RAN-compliant solution that supports all 7.x split functions defined in O-RAN. Qualcomm Technologies and Vodafone cooperate in development of 5G Open RAN Infrastructure solutions powered by the Qualcomm's X100 accelerator card and the QRU100 RAN Platform [35].
- Intel
  Intel's vRAN Dedicated Accelerator ACC100 [36] is a PCIe card supporting RAN L1 Forward Error Correction (FEC) functionality, including LDPC encoder/decoder, Code block CRC generation/checking, Rate matching/de-matching, and HARQ buffer management.
- Marvell
  Marvell has launched OCTEON 10 Fusion family of baseband processors for 5G base stations based on 5 nm process technology, which was adopted by Nokia, Samsung, and Dell [37]. In addition to multiple CPU cores and DSP cores, the chip contains RAN HW accelerators for 5G and Open RAN technology, such as Massive MIMO.

- NVIDIA
  In September 2023, NVIDIA announced deployment of the world's first GPU-Accelerated 5G Open RAN solution in the NTT DOCOMO network [38]. The solution is based on NVIDIA Aerial platform, which contains NVIDIA Aerial vRAN stack for 5G, AI frameworks, and accelerated compute infrastructure.

4.2.3 Interoperability and Testing

As interoperability in Open RAN requires more efforts than in "classic" RAN, the ecosystem members cooperate to promote interoperation and integration of Open RAN equipment from different vendors. To reach these goals, O-RAN ALLIANCE members put a lot of efforts in the PlugFests and Open Testing and Integration Centers (OTICs); see Chapter 3 and [39]. Similar activities are performed in US ACCoRD consortium [40] by AT&T, Verizon, DOCOMO, Ericsson, Nokia, Samsung, Fujitsu, Mavenir, Radisys, Rakuten, and other ecosystem members. In addition, Open RAN operators and vendors are widely involved in less formal cooperation groups. For example, Vodafone and Deutsche Telekom created Open RAN labs where multiple companies cooperate toward Open RAN interoperability.

4.3 Deployments

The White Paper [8] published in February 2023 by the MoU Group of operators, estimated current state of the global Open RAN deployments as "reaching thousands of sites." It was noted that new operators in greenfield deployments prevail. The White Paper estimated that in Europe, large-scale deployments would start in 2025, while small-scale deployments are already ongoing. Other noted focus areas are O-RAN certification, system integration, and security. For security, the publication requested that O-RAN should endorse the ENISA certification scheme for 5G and the GSMA NESAS security assurance scheme. See Chapter 3 for overview of Open RAN trials and deployments.

For the US and worldwide, the most significant event definitely is the announcement by AT&T of planned Open RAN deployment in December 2023 [12]. The plan is to start deployment of Open RAN sites with Ericsson and Fujitsu in 2024, with the target of 70% of wireless traffic going via Open RAN capable network nodes by the end of 2026. The announcement also pointed to several other vendors, Corning, Dell, and Intel that would be involved in scaling of the Open RAN share throughout the AT&T's RAN.

References

1 Deutsche Telekom, Orange, Telefonica, TIM, and Vodafone. (2021). building an Open RAN ecosystem for Europe to lead in this essential innovation.
2 Deutsche Telekom, Orange, Telefonica, TIM, and Vodafone. (2021). Memorandum of Understanding on the Implementation of Open RAN Based Networks in Europe.
3 Deutsche Telekom, Orange, Telefonica, TIM, and Vodafone. (2022). Notes on the Open RAN Technical Priorities (Release 2).
4 Deutsche Telekom, Orange, Telefonica, TIM, and Vodafone. (2022). Open RAN Technical Priority "Release 2" Document.
5 Deutsche Telekom, Orange, Telefonica, TIM, and Vodafone. (2023). Notes on the Technical Priorities (Release 3).
6 Deutsche Telekom, Orange, Telefonica, TIM, and Vodafone. (2023). Open RAN Technical Priority Document Release 3.
7 O-RAN ALLIANCE. (2023). O-RAN Release 3 Implements Features with 58 New or Updated Technical Documents. O-RAN ALLIANCE blog (March 15).
8 Deutsche Telekom, Orange, Telefonica, TIM, and Vodafone. (2023). Open RAN MoU progress update on maturity, security and energy efficiency. White Paper.
9 Ericsson. (2023). Ericsson takes leadership in industrializing Open RAN. Press Release (September 25)
10 Ericsson. (2022). What's next for RAN? Cloud RAN innovation and purpose-built evolution. White Paper.
11 Ericsson. (2024). Enhanced Ericsson portfolio focuses on performance, Sustainability and Open RAN. Press Release (February 15).
12 AT&T. (2023). AT&T to Accelerate Open and Interoperable Radio Access Networks (RAN) in the United States through new collaboration with Ericsson. AT&T News (December 04).
13 Ericsson. (2024). Ericsson and O2 Telefonica begin Open RAN Journey with first Cloud RAN deployment in Europe. Press Release (February 2024).
14 Nokia. (2023). Nokia launches anyRAN to drive CloudRAN partnerships enabling flexibility for mobile network operators and enterprises. MWC23. Press release (26 February).
15 Le Maistre, R. (2023). DT kickstarts its Open RAN rollout with Nokia, Fujitsu.
16 Nokia. (2023). Nokia Cloud RAN and Open RAN Solution Roadmap. White Paper.
17 Nokia. (2023). Cloud RAN: A Guide to Acceleration Options. White Paper.
18 Nokia. O-RAN compliant RAN programmability with Near-Real-Time RIC. https://www.nokia.com/networks/mobile-networks/mantaray-ric/ (accessed 9 March 2024).

19 Samsung. (2023). Samsung Announces the Next Phase of its 5G vRAN. Samsung Newsroom (February 27).

20 Samsung. (2023). Vodafone and Samsung Begin Mass Open RAN Rollout Across the United Kingdom. Samsung Newsroom (August 31).

21 Samsung. (2023). Samsung and Vodafone Continue To Lead Open RAN Innovation Across Europe. Samsung Newsroom (February 27).

22 Samsung. (2023). DISH Wireless Launches Virtual Open RAN 5G Network with Samsung. Samsung Newsroom (22 February).

23 Samsung. (2024). Samsung, Verizon and Keysight Advance Commercial Open RAN with Extensive Conformance and Interoperability Testing. Samsung Newsroom (February 13).

24 TELUS. (2024). TELUS partners with Samsung to build Canada's first 5G Virtualized RAN, Open RAN network. Press Release (February 15).

25 Bicheno, S. 1&1 claims launch of the first European Open RAN network. https://www.telecoms.com/open-ran/1-1-claims-launch-of-the-first-european-open-ran-network (accessed 9 March 2024)

26 Le Maistre, R. (2021). German operator 1&1 to build greenfield Open RAN network with Rakuten.

27 Le Maistre, R. (2022). Rakuten to host Open RAN test lab in UK.

28 NEC. Network Products. https://www.nec.com/en/global/solutions/network.html (accessed 30 May 2024)

29 Fujitsu. (2023). Fujitsu delivers O-RAN ALLIANCE-compliant 5G virtualized RAN solution for NTT DOCOMO's 5G commercial network services. Press Release (September 27).

30 Vodafone. (2021). Vodafone selects key partners to build Europe's first commercial Open RAN network. Vodafone News (14 June).

31 Vodafone. (2022). Vodafone deploys OpenRAN in urban locations in European first. Press Release (06 December)

32 Mavenir. (2023). DISH Wireless Expands Cloud-Native Open RAN Network with Mavenir Open vRAN Software Solutions. Press Release (23 February).

33 Khalidi, Y. (2022). Microsoft Innovation in RAN Analytics and Control. Azure Microsoft blog (December 21).

34 Qualcomm. (2022). Qualcomm Builds Momentum for Full-Scale Open RAN Commercialization with the Sampling of its 5G RAN Platforms. Press Release (28 September).

35 Vodafone. (2022). Qualcomm and Vodafone Partnering to Develop and Test Next-Generation 5G Open RAN Infrastructure. Vodafone News (25 October).

36 Intel Product Brief. (2021). Intel vRAN Accelerator ACC100 Adapter.

37 Marvell. (2023). Marvell Unveils OCTEON 10 Fusion Processor Family to Advance 5G Networks Industry-First 5 nm Baseband Processors Enable Best-in-Class Performance and Energy Efficiency for Integrated and Virtualized RAN. Marvell Newsroom (February 23).

38 Obiodu, E., Chong, C., and Noda, M. (2023). Enabling the World's First GPU-Accelerated 5G Open RAN for NTT DOCOMO with NVIDIA Aerial. NVIDIA Technical Blog (26 September).

39 O-RAN ALLIANCE. (2023). Overview of Open Testing and Integration Centre (OTIC) and O-RAN Certification and Badging Program. White Paper.

40 Soni, R. (2024). The Next Step in Bringing Our Open RAN Vision to Life. AT&T Blog (February 12).

5

Open RAN Architecture

CONTENTS

Acronyms and Abbreviations

AI	Artificial Intelligence
AI/ML	Artificial Intelligence/Machine Learning
CA	Carrier Aggregation
CN	Core Network
C-SON	Centralized SON
CU	RAN Central Unit
DC	Dual Connectivity
DL	Downlink
DRB	Data Radio Bearer
D-SON	Distributed SON
DU	RAN Distributed Unit

Open RAN: Technology and Ecosystem, First Edition. Vladimir Yanover.

eNB	LTE Base Station defined in 3GPP, eNodeB
FCAPS	Fault, Configuration, Accounting, Performance, Security management capabilities
gNB	5G Base Station defined in 3GPP, gNodeB
gNB-CU	The Control Unit of the 5G Base Station defined in 3GPP; same as gNodeB-CU
gNB-CU-CP	The Control Unit, Control Plane of the 5G Base Station defined in 3GPP; same as gNodeB-CU-CP
gNB-CU-UP	The Control Unit, User Plane of the 5G Base Station defined in 3GPP; same as gNodeB-CU-UP
gNB-DU	The Data Unit of the 5G Base Station defined in 3GPP; same as gNodeB-DU
HO	Handover
H-SON	Hybrid SON
LCM	Life Cycle Management
ML	Machine Learning
ms	millisecond
Near-RT RIC	Near-Real-Time Radio Intelligent Controller
NF	Network Function
Non-RT	Non-real time
Non-RT RIC	Non-Real-Time Radio Intelligent Controller
NRT	Non-Real Time
NSI	Network Slice Instance
NSSI	Network Slice Subnet Instance
O1	O-RAN management interface (OAM, FCAPS)
O2	O-RAN cloud management interface
O-Cloud	Cloud defined in O-RAN, which, in particular, supports O2 interface
O-CU	O-RAN defined Central Unit
O-CU-CP	O-RAN defined Central Unit – Control Plane
O-CU-UP	O-RAN defined Central Unit – User Plane
O-DU	O-RAN defined Distributed Unit
O-eNB	O-RAN conformant LTE Base Station, which, in particular, supports E2 interface
OFH	Open Fronthaul interface
O-gNB	O-RAN conformant 5G Base Station, which, in particular, supports E2 interface
O-RU	O-RAN defined Radio Unit
PM	Performance Management
PRB	Physical Resource Block in LTE and 5G
RAN	Radio Access Network

RRM Radio Resources Management
RRU Remote Radio Unit
RT Real Time
RU Radio Unit
SLA Service Level Agreement
SMO Service Management and Orchestration (framework)
SON Self-Organized Network
SST Slice/Service Type
TCO Total Cost of Ownership
TN Transport Network
UE User Equipment
UL Uplink

5.1 O-RAN Logical Architecture

O-RAN specifications mostly are focused on extension of the 3GPP 5G RAN architecture. In short, this extension is about additional split of RAN, built-in RAN cloudification, and RAN programmability. All three aspects are widely considered contributing to reduction of the RAN Total Cost of Ownership (TCO).

Additional RAN split, into components with standardized interfaces, allows entry of new vendors with component-level products. Entry of new vendors is supposed to drive down the cost of the whole system.

RAN cloudification allows RAN deployment on general-purpose cloud infrastructure instead of proprietary hardware platform, which is supposed to reduce the cost.

RAN programmability allows network operators to extend the functionality of the RAN infrastructure after it's been purchased and deployed, without additional purchase of the equipment, services, or software upgrades.

This chapter describes O-RAN Logical Architecture, defined in the O-RAN ALLIANCE specification O-RAN.WG1.OAD-R003-v11.00; see Figure 5.1.

The architecture components and interfaces, shown in Figure 5.1, are described in Sections 5.2, 5.3, 5.4, 5.5, 5.6, and 5.8. Section 5.8 outlines O-RAN High-Level Architecture.

Figure 5.1 shows the following architecture components:

- O-gNB components
  - **O-CU-CP**: the gNodeB-CU-CP defined in 3GPP, with additional requirement to support the E2 interface to the Near-RT RIC and the O1 interface to the Service Management and Orchestration (SMO)
  - **O-CU-UP**: the gNodeB-CU-UP defined in 3GPP, with additional requirement to support the E2 interface to the Near-RT RIC and the O1 interface to the SMO

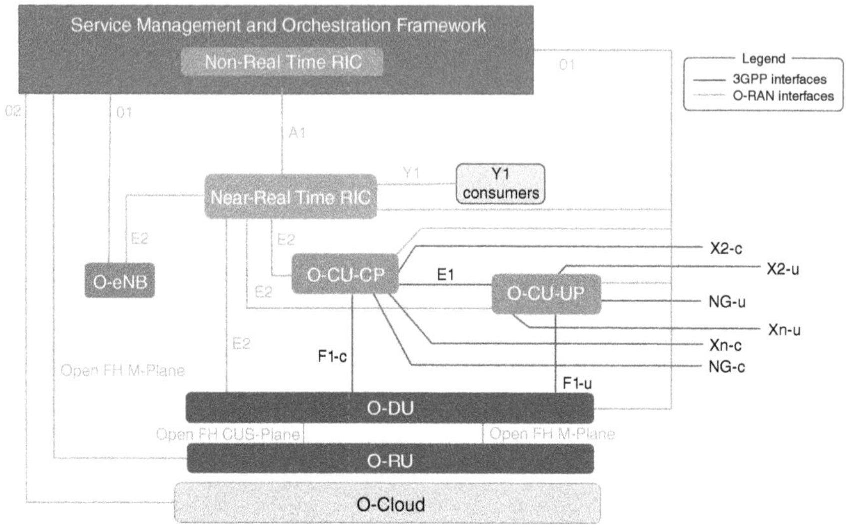

Figure 5.1 Logical Architecture of O-RAN. (*Source:* Reproduced with permission of O-RAN ALLIANCE e.V.)

- **O-DU**: the node, implementing the functions of the gNodeB-DU defined in 3GPP, except the functions moved to O-RU per the 7.2x split (see Chapter 8), and with additional requirements to support the E2 interface to the Near-RT RIC and the O1 interface to the SMO. The O-DU supports the OFH interface to the O-RU, including the M-Plane management
- **O-RU** node implements the functions of the gNodeB-DU moved to the O-RU per the 7.2x split (see Chapter 8); the O-RU supports the OFH interface to the O-DU and the M-Plane management interface to the O-DU and the SMO

- **O-eNB**: the eNodeB or ng-eNB defined in 3GPP, with additional requirement to support the E2 interface to the Near-RT RIC
- **O-Cloud**: a cloud computing platform, hosting such O-RAN components as Near-RT RIC, O-CU (O-CU-CP, O-CU-UP), and O-DU. The O-Cloud allows cloud infrastructure and Network Functions management via the O2 interface; see O-RAN.WG6.CADS-v06.00 [1] and O-RAN.WG6.ORCH-USE-CASES-R003-v09.00 [2]
- **Service Management and Orchestration (SMO) framework**: supports Fault, Configuration, Accounting, Performance, Security management capabilities (FCAPS) interface to O-RAN Network Functions, policy-based A1 interface to the Near-RT RIC, O2 interface to O-Cloud for cloud management and orchestration, and OFH M-Plane interface for management of the O-RU

- **Non-Real-Time Radio Intelligent Controller (Non-RT RIC)**: an SMO component, which is capable of non-real-time control of the Near-RT RIC via the A1 interface
- **Near-Real-Time Radio Intelligent Controller (Near-RT RIC)**: an O-RAN function, capable of control and optimization of O-RAN components in near-real time via the E2 interface.

Figure 5.1 shows the following O-RAN interfaces:

- **A1**: policy-based interface between the Non-RT RIC and the Near-RT RIC for non-real-time control and optimization (see O-RAN.WG2.A1GAP-v03.02 [3], O-RAN.WG2.A1UCR-R003-v01.02 [4])
- **O1**: management interface between the SMO framework and O-RAN managed elements, supporting OAM/FCAPS management; see O-RAN.WG10. Information Model and Data Models.0-R003-v07.00 [5], O-RAN.WG10.O1-Interface.0-R003-v12.00 [6], O-RAN.WG10.OAM-Architecture-R003-v11.00 [7].
- **Open Fronthaul (OFH)**: interface between the O-DU and the O-RU; includes Control Plane, User Plane, and Synchronization Plane (O-RAN.WG4.CUS.0-R003-v14.00 [8]), and OFH M-Plane interface for OAM management of the O-RU (see O-RAN.WG4.MP.0-R003-v14.00 [9])
- **O2**: interface for O-Cloud management between the SMO framework and the O-Cloud (see O-RAN.WG6.O2-GA&P-R003-v06.00 [10])

Additional split of RAN, built-in RAN cloudification, and RAN programmability create a strong market drive for Open RAN technology progress. With standardization of the corresponding interfaces: Open Fronthaul, E2, A1, O1, O2, the O-RAN architecture enables interoperability between RAN components, manufactured by different vendors. A vendor can offer any of the components O-CU (O-CU-CP, O-CU-UP), O-DU, O-RU rather than integrated RAN solutions. Programmability of O-RAN is based on capabilities of the SMO, Non-RT RIC, and Near-RT RIC and on support of the corresponding interfaces O1, O2, A1, E2 in the RAN components and in the O-RAN Cloud (O-Cloud). Support for third-party applications rApps running on the Non-RT RIC platform and xApps running on the Near-RT RIC platform, creates an unlimited potential for extension of the RAN intelligent capabilities.

5.1.1 Additional RAN Split

3GPP has already defined split of 5G RAN node into the Central Unit (gNodeB-CU or gNB-CU) and Distributed Unit (gNodeB-DU or gNB-DU) with possible further split of CU into the Control Plane node gNodeB-CU-CP (gNB-CU-CP) and User Plane node gNodeB-CU-UP (gNB-CU-UP). Standardization of the

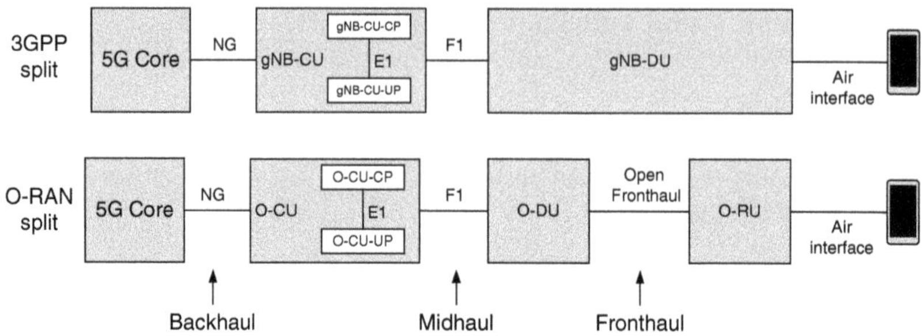

Figure 5.2 RAN Decomposition in O-RAN architecture. (*Source:* Vladimir Yanover.)

interfaces between 5G RAN components (e.g., gNodeB-CU and gNodeB-DU) enables interoperation between the components manufactured by different vendors.

Figure 5.2 shows how O-RAN takes the RAN split to the next level by further splitting the gNodeB-DU (3GPP) into the O-RAN Distributed Unit (O-DU) and O-RAN Radio Unit (O-RU) with additional Open FrontHaul (OFH) interface between them.

5.1.2 RAN Programmability

RAN programmability allows better adjustment of the RAN behavior to the local environment, resulting in better control and optimization of the network. In addition, RAN programmability enables introduction of new services, not foreseen at the time of RAN purchase and deployment. O-RAN architecture enables RAN programmability by introducing additional layer of intelligence in Service Management and Orchestration (SMO) framework, Non-RT RIC and Near-RT RIC controllers. While Non-RT RIC enables RAN programmability in non-real time, Near-RT RIC enables monitoring and control of the RAN components O-CU (O-CU-CP, O-CU-UP) and O-DU in near-real time. Non-RT RIC and Near-RT RIC are platforms, capable of hosting third-party applications. The hosted applications can access network state and performance information and issue optimization commands, which provides for additional flexibility. The applications running on Non-RT RIC and Near-RT RIC can be produced by the network operator itself or purchased from third-party vendors.

5.1.3 Integrated Cloud Management

In addition to architecture extension, O-RAN established the concept of integrated cloud (O-Cloud), where the Open RAN components can be hosted. As can be seen in Figure 5.1, in the O-RAN architecture, the SMO is a central manager, which can manage both RAN components and the underlying O-Cloud. The O1 interface enables FCAPS management of RAN components. The A1 interface enables policy management of the Near-RT RIC. The O2 interface enables management of the cloud where all or some of O-RAN network functions are hosted (O-Cloud).

5.2 Service Management and Orchestration (SMO)

Service Management and Orchestration (SMO) framework [11] is responsible for management of O-RAN Network Functions. This includes classic OAM (FCAPS) network management system capabilities per 3GPP 32.xxx and 28.xxx specification series, but also new functionalities, such as cloud management and non-real-time policy-based RAN control and optimization. SMO supports the following interfaces:

- O1 (OAM) interface, which enables Fault, Configuration, Accounting, Performance, Security (FCAPS) management capabilities toward O-RAN Network Functions, such as Near-RT RIC, O-CU (O-CU-CP, O-CU-UP), O-DU
- Open FH M-Plane interface for O-RU management
- O2 interface for O-Cloud management and orchestration of the cloud resources

The FCAPS functions supported by the SMO are described in OAM architecture specification O-RAN.WG10.OAM-Architecture-R003-v11.00 [7] and O1 interface specifications O-RAN.WG10.O1-Interface.0-R003-v12.00 [6]. The O1 interface mostly follows 3GPP specifications for Configuration Management (CM), Performance Management (PM), Fault Management (FM), and Trace. In addition to that, the O1 interface supports File Management, Heartbeat service, and the functions related to Physical Network Functions (PNFs): PNF Discovery and PNF Software Management.

SMO includes Non-Real-Time Radio Intelligent Controller (Non-RT RIC) as a part (Figure 5.3). Non-RT RIC is used for control of the RAN Network Functions and of the Near-RT RIC. There is no definition of the interface between the Non-RT RIC and other SMO components.

The Non-RT RIC can use the SMO functionalities. For example, Non-RT RIC can request other SMO components to collect specific network performance measurements via the O1 interface. After the collection of the measurements is completed, the Non-RT RIC can retrieve the collected information; see Chapter 9.

5.3 Non-RT RIC: Non-Real-Time RAN Intelligent Controller

Non-RT RIC (Figure 5.3) is a function, internal to the SMO, used for intelligent non-real-time RAN control and optimization (see O-RAN.WG1.OAD-R003-v11.00 [11], O-RAN.WG2.Use-Case-Requirements-v09.00 [12], O-RAN.

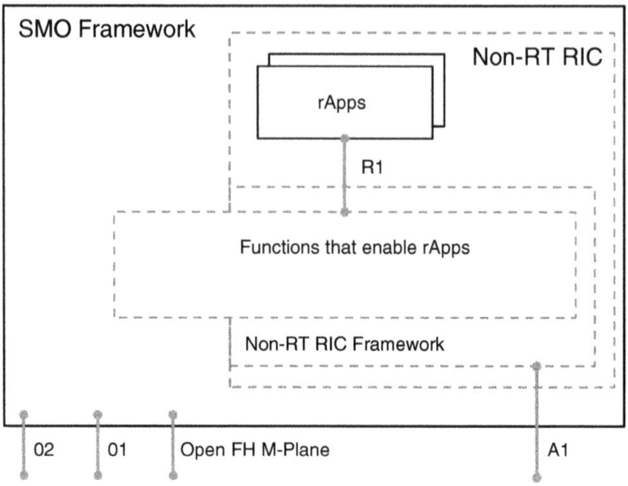

Figure 5.3 SMO framework and Non-RT RIC. (*Source:* Reproduced with permission of O-RAN ALLIANCE e.V.)

WG2.Non-RT-RIC-ARCH-R003-v05.00 [13]). For this purpose, the Non-RT RIC provides policy-based guidance and supplies Enrichment Information to the Near-RT RIC. The optimization goals can be set, for example, in terms of Radio Resources Management (RRM) in the time scale of seconds or more, termed non-real time. For data collection, the Non-RT RIC can use SMO management services. For optimization actions, Non-RT RIC can use provisioning services, consumed by the SMO; these actions can be directed toward the following O-RAN components:

- O-RAN Network Functions, via O1 interface
- O-Cloud, via O2 interface
- O-RU via OFH M-Plane interface
- Near-RT RIC by setting A1 policies via the A1 interface

To determine necessary optimization actions, Non-RT RIC can use AI/ML training/inference. In addition, Non-RT RIC provides ML model management services; see Chapter 11.

Figure 5.4 shows internal architecture of the Non-RT RIC as a hosting platform for third-party applications, called rApps. The rApp applications implement RAN control/optimization algorithms and may implement other functions, such as management of database where network information is stored. The hosting platform provides rApps with access to management services consumed by the SMO, such as Performance Management (PM), Configuration Management (CM), and A1 policy management. The rApps may implement AI/ML services exposed to other rApps, to Non-RT RIC, Near-RT RIC, and other RAN components.

The component called "Non-RT RIC Framework" provides necessary support to rApps via the R1 interface (O-RAN.WG2.R1GAP-v07.00 [14], O-RAN.WG2.R1UCR-R003-v06.00 [15]). This component provides to rApps support for collection of OAM-level network performance data (O1) and for non-real-time network control and optimization actions via the A1, O1, O2, and Open FH M-Plane interfaces.

More details on SMO and Non-RT RIC can be found in Chapter 6.

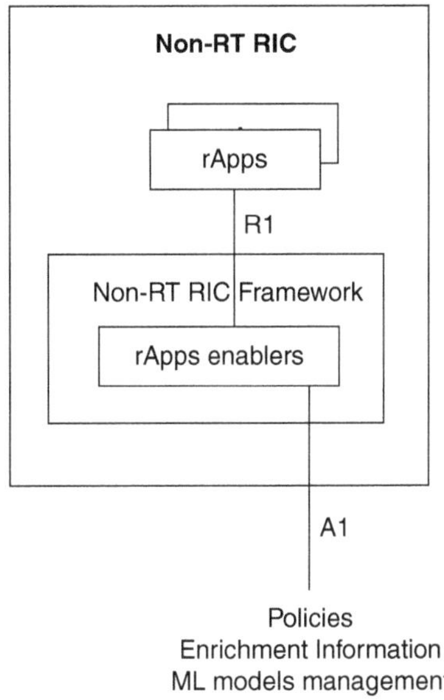

Figure 5.4 Non-RT RIC. (*Source:* Vladimir Yanover.)

5.4 Near-RT RIC: Near-Real-Time RAN Intelligent Controller

Near-RT RIC is a logical function, used for control and optimization of RAN components in near-real time (O-RAN.WG3.RICARCH-R003-v05.00 [16]). Non-RT RIC provides to the Near-RT RIC a guidance for RAN control and optimization in the form of "A1 Policies" via the A1 interface. The policies, in particular, may set the optimization goals. In addition to policies, the Near-RT RIC can receive from the Non-RT RIC data from external sources, called Enrichment Information (EI).

Figure 5.5 shows the architecture of Near-RT RIC with its connections to RAN components ("E2 Nodes") via the E2 interface. The O-RAN network functions supporting E2 interface: O-CU, O-CU-CP, O-CU-UP, O-DU, or any combination (integration) of those. For the E2 interface, O-RAN specifications define E2 Services. E2 Nodes play the role of service providers, while the Near-RT RIC plays the role of service consumer.

E2 Services support near-real-time data collection and control, which can be used for optimization. The specifications O-RAN.WG3.E2GAP-R003-v05.00 [17] and O-RAN.WG3.E2AP-R003-v05.00 [18] define E2 Application Part protocol used for this purpose. With E2 Services, Near-RT RIC can configure an E2 Node for reporting real-time performance information and/or for notification on the events, which occur in the E2 Node. After the Near-RT RIC receives this information, it can respond in near-real time with action steering the E2 Node's behavior to the desired direction. For that, the Near-RT RIC can re-configure ongoing event-handling procedure in the E2 Node or trigger different event handling procedure. Alternatively, the Near-RT RIC can program the E2 Node for desired

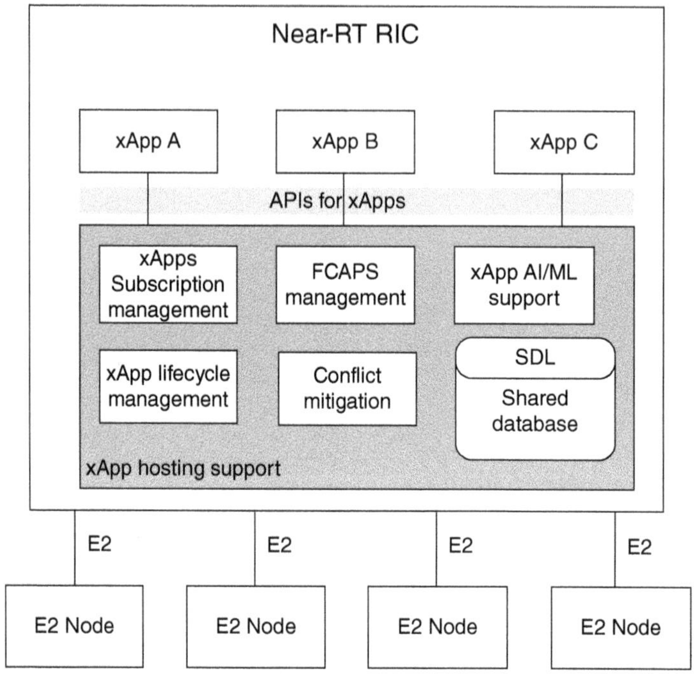

Figure 5.5 Near-RT RIC Internal Architecture. (*Source:* Vladimir Yanover.)

action in response to specific conditions and/or events. This kind of programming is performed by setting the E2 policies via the E2 interface. In such case, processing of the event is executed locally in the E2 Node, which in general allows faster response.

For the E2 Services, O-RAN specifications define universal operations (procedures, services), which can be used with any E2 Node. Separately from that, the specifications define several information models, called "E2 Service Models" (E2SMs), which enable support of services specific for certain type of E2 Nodes or specific functionalities. See a general description of the E2SM technique in O-RAN.WG3.E2SM-R003-v05.00 [19] and specific E2SMs in O-RAN.WG3.E2SM-CCC-R003-v03.00 [20], O-RAN.WG3.E2SM-KPM-R003-v04.00 [21], and O-RAN.WG3.E2SM-RC-R003-v05.00 [22].

Figure 5.5 shows the architecture of the Near-RT RIC as a platform hosting third-party applications called xApps. The host provides xApps with access to E2 Services for collection of network data and RAN control, which allows implementation of RAN control and optimization algorithms in the xApps; see O-RAN-WG3.RICAPI-R003-v01.00 [23]. This part of the O-RAN architecture essentially extends capabilities related to RAN programmability. Additionally, the host provides xApps with access to the subscription service, to E2 Nodes FCAPS management, and to the RAN information in the shared database. There is some level of support for conflict mitigation and xApp life cycle management.

More details on Near-RT RIC and E2 interface can be found in Chapter 7.

5.5 RAN Nodes and RAN Components

O-gNB is a 5G RAN gNB (gNodeB) network function (see 3GPP TS 38.401 [24]), which supports additional functionalities defined in O-RAN and the corresponding interfaces E2, O1. As specified in O-RAN, O-gNB may be split into several components, such as O-CU (with further split into O-CU-CP, O-CU-UP), O-DU, and O-RU. The components support the interfaces N2, N3, Xn, X2 defined in 3GPP. In addition, the O-gNB components are supposed to support E2 interface (to be "E2 Nodes") and provide the corresponding services ("E2 Services") to Near-RT RIC. Information of RAN split chosen in O-RAN ALLIANCE can be found in Chapter 8. O-DU and O-RU should support Open Fronthaul interface defined in O-RAN.WG4.CUS.0-R003-v14.00 [8]. In the part of network management, the O-gNB component should support O1 management interface, which is aligned to management interfaces and services, defined in the 28.xxx and 32.xxx families of 3GPP specifications

O-eNB is defined as an eNB (3GPP TS 36.401 [25]) or ng-eNB (3GPP TS 38.401 [24]) that supports the E2 interface.

5.6 O-Cloud

The term "O-Cloud" means a cloud platform customized for virtualized Open RAN, i.e. hardware and software that provide computing, networking, and storage capabilities for the RAN network functions hosted by the cloud. In addition to general-purpose compute, networking, and storage infrastructure components, O-Cloud may also include various RAN-specific acceleration devices. Such accelerators allow to achieve required performance objectives at reasonable cost. O-Cloud is required to support the O-RAN O2 interface and expose open APIs enabling the life cycle management of virtualized network functions on the cloud.

More details on O-Cloud can be found in Chapter 8.

5.7 Reference Points (Interfaces)

The architecture depicted in Figure 5.1 identifies the following reference points (interfaces):

- A1 Interface between the SMO/Non-RT RIC and the Near-RT RIC for non-real-time RAN control and optimization. The A1 interface allows Non-RT RIC to set the policies in the Near-RT RIC, receive policy feedback, and deliver Enrichment Information (EI) to the Near-RT RIC

- E2 interface between the Near-RT RIC and a RAN component (E2 Node), such as O-gNB, O-CU, O-CU-CP, O-CU-UP, O-DU, or any combination of those. This interface enables a rich set of E2 Services, supplied by E2 Nodes to the Near-RT RIC. The services allow:
  - Subscription for delivery of network performance data in near-real time from the E2 Nodes, with granularity level of network nodes, cells, slices, individual UEs, and individual QoS Flows
  - Subscription for near-real-time notifications on the events in E2 Nodes, with granularity level of events related to network nodes, cells, slices, individual UEs, individual QoS Flows, and transfer of individual protocol messages over RAN interfaces
  - Sending near-real-time control commands to E2 Nodes
  - Setting policies in the E2 Nodes that prescribe reaction of the E2 Node to internal events
- O1 Interface between the SMO framework and O-RAN managed elements, for OAM/FCAPS management. For the O1 interface, the management services are defined in line with the corresponding 3GPP network management specifications. The O1 management information models essentially are extensions of the corresponding 3GPP management information models
- Open FH M-Plane interface used by the SMO and/or O-DU for OAM management of the O-RU
- O2 Interface between the SMO framework and the O-Cloud for management of O-Cloud
- Y1 interface used for exposure of the analytics data generated by the Near-RT RIC

5.8 O-RAN High-Level Architecture

Figure 5.6 depicts O-RAN High-Level Architecture (O-RAN.WG1.OAD-R003-v11.00 [11]), which can be derived from "full" O-RAN Logical Architecture in Figure 5.1 by combining Near-RT RIC and O-gNB in a single entity (block) called "O-RAN Network Functions" and collapsing the E2 interface. In fact, it depicts a special deployment model, where the functionalities of O-gNB and Near-RT RIC are either integrated or connected via proprietary interface.

The High-Level Architecture enables interoperability:

- Between the SMO/Non-RT RIC and combined "O-RAN Network Functions" block (A1, O1, and OFH M-Plane interfaces)
- Between the SMO/Non-RT RIC and O-Cloud (O2 interface)

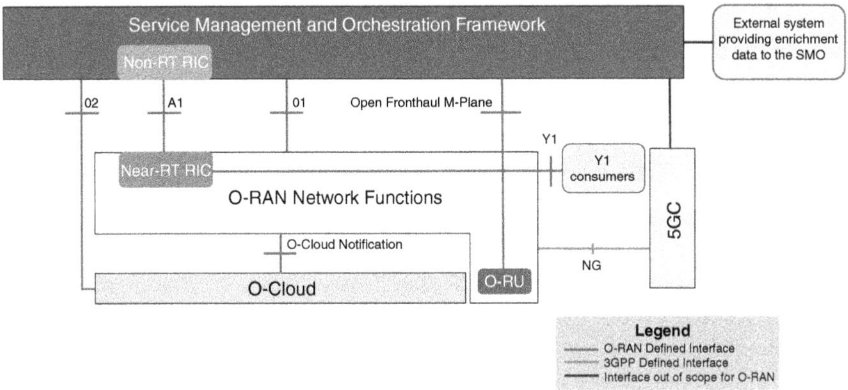

Figure 5.6 High-Level Architecture of O-RAN. (*Source:* Reproduced with permission of O-RAN ALLIANCE e.V.)

References

1 O-RAN.WG6.CADS-v06.00. Cloud Architecture and Deployment Scenarios for O-RAN Virtualized RAN
2 O-RAN.WG6.ORCH-USE-CASES-R003-v09.00. Cloudification and Orchestration Use Cases and Requirements for O-RAN Virtualized RAN
3 O-RAN.WG2.A1GAP-v03.02. O-RAN A1 Interface: General Aspects and Principles
4 O-RAN.WG2.A1UCR-R003-v01.02. A1 Interface: Use Cases and Requirements
5 O-RAN.WG10. Information Model and Data Models.0-R003-v07.00. O-RAN Information Model and Data Models Specification
6 O-RAN.WG10.O1-Interface.0-R003-v12.00. O-RAN Operations and Maintenance Interface Specification
7 O-RAN.WG10.OAM-Architecture-R003-v11.00. O-RAN Operations and Maintenance Architecture
8 O-RAN.WG4.CUS.0-R003-v14.00. O-RAN Control, User and Synchronization Plane Specification
9 O-RAN.WG4.MP.0-R003-v14.00. O-RAN Management Plane Specification
10 O-RAN.WG6.O2-GA&P-R003-v06.00. O-RAN O2 Interface General Aspects and Principles 6.0
11 O-RAN.WG1.OAD-R003-v11.00. O-RAN Architecture Description
12 O-RAN.WG2.Use-Case-Requirements-v09.00. Non-RT RIC & A1 Interface: Use Cases and Requirements
13 O-RAN.WG2.Non-RT-RIC-ARCH-R003-v05.00. Non-RT RIC Architecture
14 O-RAN.WG2.R1GAP-v07.00. O-RAN R1 Interface: General Aspects and Principles

15 O-RAN.WG2.R1UCR-R003-v06.00. O-RAN R1 Interface: Use Cases and Requirements

16 O-RAN.WG3.RICARCH-R003-v05.00. Near-RT RIC Architecture

17 O-RAN.WG3.E2GAP-R003-v05.00. Near-Real-Time RAN Intelligent Controller and E2 Interface. E2 General Aspects and Principles (E2GAP)

18 O-RAN.WG3.E2AP-R003-v05.00. Near-Real-Time RAN Intelligent Controller: Use Cases and Requirements

19 O-RAN.WG3.E2SM-R003-v05.00. Near-Real-Time RAN Intelligent Controller and E2 Interface. E2 Service Model (E2SM)

20 O-RAN.WG3.E2SM-CCC-R003-v03.00. Near-Real-Time RAN Intelligent Controller. E2 Service Model (E2SM), Cell Configuration and Control

21 O-RAN.WG3.E2SM-KPM-R003-v04.00. Near-Real-Time RAN Intelligent Controller. E2 Service Model (E2SM) KPM

22 O-RAN.WG3.E2SM-RC-R003-v05.00. O-RAN E2 Service Model (E2SM), RAN Control

23 O-RAN-WG3.RICAPI-R003-v01.00. Near-RT RIC APIs specification

24 3GPP TS 38.401. NG-RAN; Architecture description

25 3GPP TS 36.401. E-UTRAN; Architecture description

6

Non-Real-Time RAN Programming

CONTENTS

Open RAN: Technology and Ecosystem, First Edition. Vladimir Yanover.

Acronyms and Abbreviations

A1-P	A1 Policy service
A1-EI	A1 Enhancement Information service
A1AP	A1 Application Protocol
DMS	Deployment Management Services
E2AP	E2 Application Protocol
E2SM	E2 Service Model
EI	A1 Enrichment Information
FCAPS	Fault, Configuration, Accounting, Performance, Security management functions
gNB-CU	gNB – Central Unit
GAD	Geographical Area Description
gNB-CU-CP	gNB – Central Unit – Control Plane
gNB-CU-UP	gNB – Central Unit – User Plane
gNB-DU	gNB – Distributed Unit
HTTP	HyperText Transfer Protocol
IE	Information Element
IMS	Infrastructure Management Services
KPM	Key Performance Measurements
MOS	Mean Opinion Score, metric to measure the overall voice or video call quality as perceived by human
NSI	Network Slice Instance
NSSI	Network Slice Subnet Instance
Near-RT RIC	Near-Real-Time Radio Intelligent Controller
NF	Network Function
NFO	Network Function Orchestration
Non-RT RIC	Non-Real-Time Radio Intelligent Controller
O-Cloud	A cloud platform, which includes a set of hardware and software components that provide cloud computing capabilities to execute O-RAN network functions; it exposes defined in O-RAN APIs enabling management of the entire life cycle for network functions
O-CU	O-RAN Central Unit
O-CU-CP	O-RAN Central Unit – Control Plane
O-CU-UP	O-RAN Central Unit – User Plane
O-DU	O-RAN Distributed Unit
O-RU	O-RAN Radio Unit
OAM	Operations, Administration, and Management
PNF	Physical Network Function
QoS	Quality of Service

QoE	Quality of Experience, metrics to measure end-to-end network service quality from the perspective of a user
REST	REpresentational State Transfer protocol
rApp	Third-party application hosted by the Non-RT RIC
SMO	Service Management and Orchestration platform
TA	Tracking Area
TAI	Tracking Area Identity
TCO	Total Cost of Ownership
VNF	Virtualized Network Function
xApp	Third-party application hosted by the Near-RT RIC

6.1 Introduction

RAN programmability is considered an essential driving power of O-RAN technology. One of the building blocks of RAN programmability is based on capabilities of Service Management and Orchestration framework (SMO) and Non-Real-Time Radio Intelligent Controller ("Non-RT RIC").

In the O-RAN architecture (Chapter 5), SMO is responsible for network management (O-RAN.WG1.OAD-R003-v11.00 [1]). This includes classic FCAPS (Fault, Configuration, Accounting, Performance, Security) management capabilities per 3GPP 32.xxx and 28.xxx specification series, but also new functionalities, such as cloud management and non-real-time, policy-based RAN control and optimization. SMO supports the interfaces O1, O2, and A1.

SMO includes Non-Real Time Radio Intelligent Controller (Non-RT RIC), where A1 protocol is terminated. Non-RT RIC is also a platform hosting third-party application ("rApps"); the functions supporting rApps are placed in the SMO either inside or outside of the Non-RT RIC (O-RAN.WG2. Non-RT-RIC-ARCH-R003-v05.00 [2]).

Network operators believe (e.g. Vodafone and NTT DOCOMO [3]) that, along with RAN TCO reduction, network programmability with SMO and Non-RT RIC can improve service availability. Example: SMO can manage automatic redeployment of RAN NFs to other locations (network healing) in case a part of the infrastructure is disrupted. Another example is SMO support for automated intelligent configuration management, with handling of numerous 5G radio parameters. Use of rApps allows deep modification of the network behavior without need in upgrade of the RAN software and firmware.

This chapter describes the architecture of the SMO framework and Non-RT RIC, and A1-based aspects of RAN programmability. The aspects based on O1 and O2 interfaces operations are addressed in Chapters 10 and 8 correspondingly.

Examples of RAN automation and optimization scenarios, where SMO and Non-RT RIC are involved, can be found in Chapter 9.

6.2 SMO and Non-RT RIC in O-RAN Architecture

6.2.1 Service Management and Orchestration Framework (SMO)

In the O-RAN architecture, the SMO is responsible for RAN management (O-RAN.WG1.OAD-R003-v11.00 [1]):

- FCAPS (OAM) interface to O-RAN Network Functions via O1 interface
- Open Fronthaul M-Plane interface to the O-RU for FCAPS support (in so-called hybrid model of O-RU management).
- Policy-based management of the Near-Real-Time Radio Intelligent Controller (Near-RT RIC) via A1 interface
- O-Cloud Management and orchestration, workflow management via O2 interface

Figure 6.1 shows the structure of SMO, including third-party applications rApps, hosted by the Non-RT RIC. The rApps enablers provide necessary support for rApps; the enablers may be placed inside or outside the Non-RT RIC.

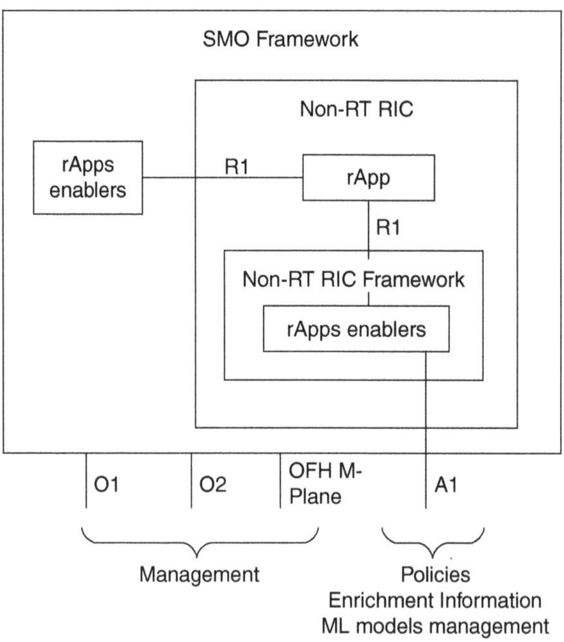

Figure 6.1 SMO Framework and Non-RT RIC. (*Source:* Vladimir Yanover.)

6.2.2 Non-Real-Time Radio Intelligent Controller (Non-RT RIC)

The SMO contains Non-RT RIC as a functional part terminating the A1 interface; see Figure 6.1 and O-RAN.WG1.OAD-R003-v11.00 [1]. The Non-RT RIC is driving intelligent RAN optimization in non-real-time domain by setting A1 policies to the Near-RT RIC. The Non-RT RIC can be also involved in AI/ML models management: model training and non-real-time inference. The A1 interface can be also used for supplying Enrichment Information to the Near-RT RIC.

Interface between the Non-RT RIC and the rest of the SMO is not defined. However, the O-RAN specifications assume that the Non-RT RIC has access to network performance data collected by the SMO via O1 interface and can initiate Configuration Management (CM) operations via the O1 interface. Similarly, the Non-RT RIC has access to information collected by the SMO via O2 interface and can trigger actions via O2 interface, such as orchestration operations.

6.2.3 rApps Hosted in Non-RT RIC; R1 Interface

The Non-RT RIC Framework exposes the R1 interface; see Figure 6.1, O-RAN.WG2.R1GAP-v07.00 [4], and O-RAN.WG2.R1UCR-R003-v06.00 [5]. The services consumed via R1 interface are produced in the part denoted "rApps Enablers." The SMO may contain additional enablers of this kind, which are not a part of the Non-RT RIC.

An rApp can play the role of service producer and/or of the service consumer. Exposure of the services normally is associated with use of open software APIs. A producer of the service can register as such; then potential consumers of the service use the registration information for discovery of the producers. To get an access to the service, a potential consumer may need to use the R1 interface for authentication and authorization. Examples of such services: the services that provide access to specific network information, the services that allow communication between rApps, the services that allow communication between an rApp and Near-RT RIC via A1 interface, etc.

The set of services, which can be accessed via the R1 interface includes the following categories:

- Service management and exposure, such as service registration and service discovery
- Data management and exposure services, which can be used by Data Consumers for delivery of network performance data from Data Producers
- A1-related services, which provide access to functionalities associated with use of A1 interface, such as A1 policy management service (Section 6.3.3) and A1 Enrichment Information service (Section 6.4)

- AI/ML workflow services (Chapter 11)
- RAN OAM-related services, which provide access to management interfaces of RAN nodes via O1, with special case of O-RU management via Open Fronthaul M-plane interface
- O2-based cloud management services (Chapter 8)
- rApp management services for configuration of rApps, rApps performance reporting, rApps faults reporting, rApps log information generation and delivery.

6.3 A1 Interface

6.3.1 Introduction

This section describes basics of the A1 interface: the concept, architecture, services, operations, and procedures. Description of these aspects is accompanied with examples of A1 REST realization adopted in O-RAN.

The section is based on the following O-RAN specifications

- **O-RAN.WG2.Use-Case-Requirements-v09.00**: Non-RT RIC & A1 Interface: Use Cases and Requirements [6] describes the targeted use cases
- **O-RAN.WG2.A1GAP-R003-v03.02, A1 interface**: General Aspects and Principles [7]
- **O-RAN.WG2.A1AP-R003-v04.01, A1 interface**: Application Protocol [8]. This specification contains the definition of the A1 Application Protocol (A1AP), which follows the Network Function Service Framework specified in TS 23.501 [9], Section 7.1. The O-RAN specification describes the REST realization of the A1 interface, based on the design principles defined in 3GPP TS 29.501 [10] and the design patterns for HTTP/JSON, defined in TS 32.158 [11].
- **O-RAN.WG2.A1TD-v07.00, A1 interface**: Type Definitions [12], defines the information model of the A1 interface

Figure 6.2 shows typical network control/optimization scenario, where Non-RT RIC receives via OAM the high-level goals (intents) and monitors the Performance Management (PM) information ("counters") received by the SMO via the O1 interface. Another source of information can be A1 policy feedback, where the Near-RT RIC notifies of changing the policy status, e.g. from ENFORCED to NOT_ENFORCED. Based on these inputs, the Non-RT RIC determines the actions needed to reach the high-level goals. Execution of the actions includes (re)configuration of the Near-RT RIC and RAN components O-gNB/O-CU/O-CU-CP/O-CU-UP/O-DU. In addition, Non-RT RIC can act toward the Near-RT RIC by setting or modifying A1 Policies. For example, such action may be to modify the A1 Policy in the part of QoS targets for certain category of UEs [6].

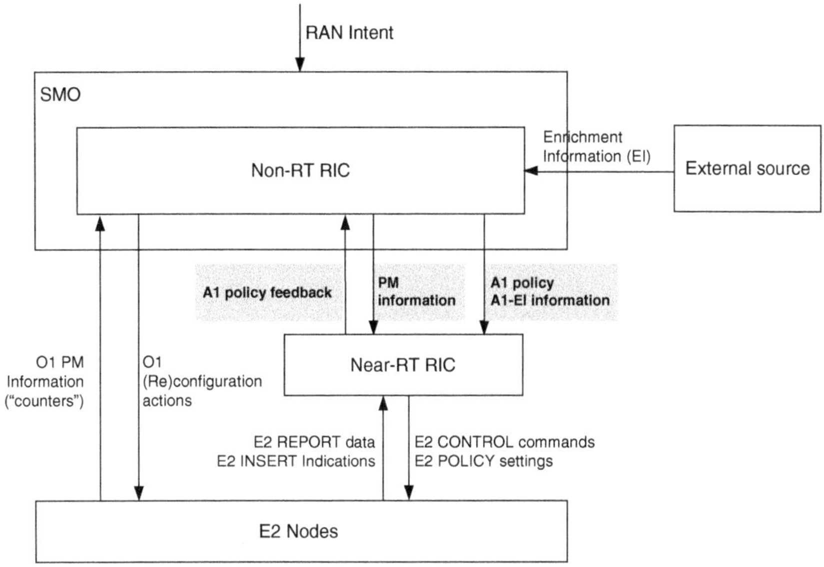

Figure 6.2 A1 interface in typical O-RAN control/optimization scenario. (*Source:* Vladimir Yanover.)

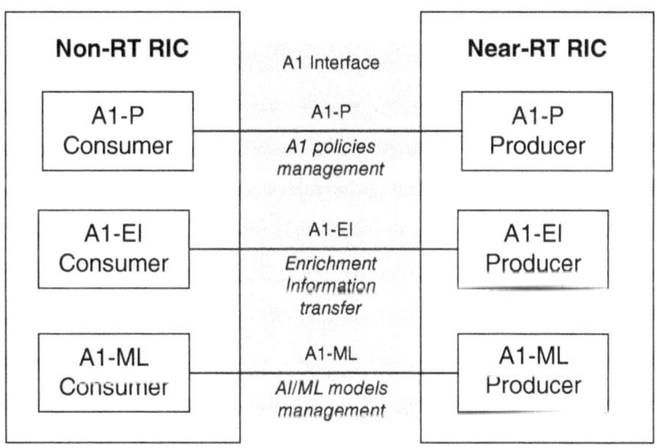

Figure 6.3 A1 services. (*Source:* Vladimir Yanover.)

Figure 6.3 shows A1 service interfaces: A1-P for policies (see Section 6.3.3), A1-EI for Enrichment Information (Section 6.4), and A1-ML for support of ML operations. The A1-ML service is addressed in [13].

6.3.2 Basics of A1 Interface REST Realization

Figure 6.4 shows the A1 protocol stack in the REST realization, with HTTP on top of TCP/IP (O-RAN.WG2.A1GAP-R003-v03.02 [7]). The information transferred between the Non-RT RIC and Near-RT RIC is JSON encoded.

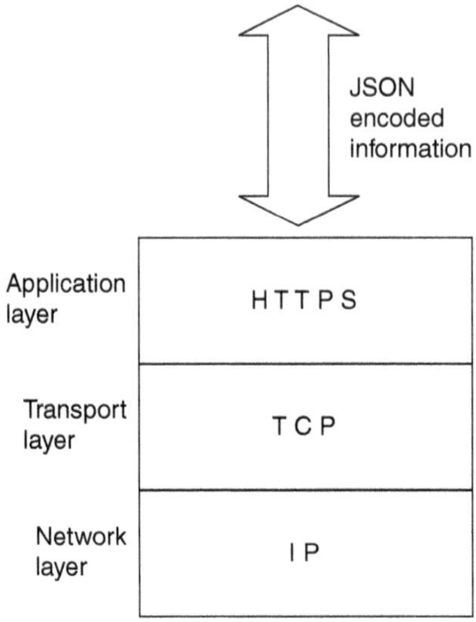

Figure 6.4 A1 Protocol Stack. (*Source:* Vladimir Yanover.)

In line with modern Service-Based approach, the A1-P and A1-EI protocol specifications describe the information model (O-RAN.WG2.A1TD-R003-v07.00 [12]) separately from protocol operations (O-RAN.WG2.A1AP-R003-v04.01 [8]).

6.3.3 A1 Policy Management Service (A1-P)

The specification of the A1 Application Protocol A1AP (O-RAN.WG2.A1AP-R003-v04.01 [8]) follows the 3GPP framework described in 3GPP TS 29.501 [10], and is aligned to the REST design rules per 3GPP TS 32.158 [11].

The A1-P service enables policy management with Near-RT RIC being the service producer and Non-RT RIC the service consumer. The A1-P service is based on the operations discussed in Section 6.3.9.

It's important to note that for every policy, the Near-RT RIC holds policy enforcement status, which normally is set to ENFORCED, but can be changed unilaterally by the Near-RT RIC to NOT_ENFORCED, in which case it should notify the Non-RT RIC, using A1 policy feedback mechanism. The Create policy request operation may return success or error indication, with corresponding consequences to the enforcement state visible at the Non-RT RIC. Some operations may be followed by or caused by a change in the enforcement status. For example, "Notify policy status change" operation is triggered by the status change.

6.3.4 The Structure of A1 Policy

An A1 policy consists of a scope statement (object, identifier) and one or more of policy statements (Figure 6.5).

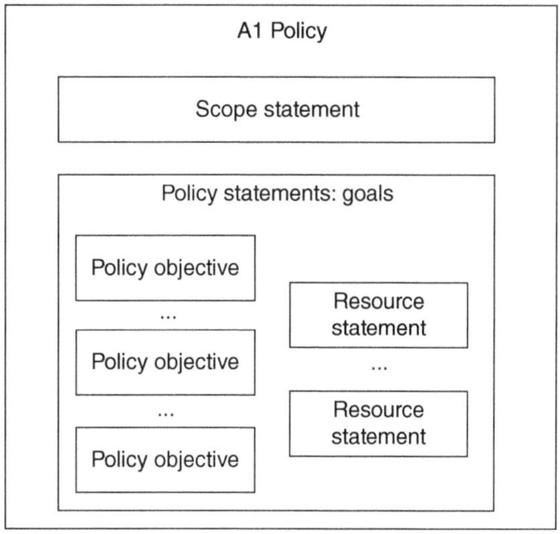

Figure 6.5 Illustration of the A1 policy structure. (*Source:* Vladimir Yanover.)

The scope statement identifies the objects, to which the policy statements are applied, such as individual UEs, UE groups, QoS Flows, network slice instances, and cells. Every policy statement sets the goals to the Near-RT RIC; it contains Resource statements representing the policy resources and Policy objectives.

The Near-RT RIC interprets the A1 policy as a guidance: within the policy scope (e.g. a group of UEs), reach the Policy objectives (e.g. UE level guaranteed bit rate), while satisfying the conditions specified in the Resource statements (e.g. prioritize use of certain cells over use of other cells).

Every A1 policy instance belongs to certain policy type supported by the Near-RT RIC. The policy type is defined by the PolicyTypeObject (O-RAN.WG2.A1AP-R003-v04.01 [8]), which contains JSON schemas to be used for validation of a PolicyObject, e.g. when the Near-RT RIC receives a Policy creation request. In such case, PolicyTypeId string identifies the policy type. At the initialization of the A1 interface, the Non-RT RIC can query the Near-RT RIC about the supported policy types; the response contains the list of supported policy types.

In the REST realization, the policies are considered HTTP resources. Figure 6.6 shows the structure of the corresponding URIs. The policy attributes are included in the PolicyObject in the HTTP message body.

Figure 6.6 specifies the structure of the resource URI for the A1-P API.

Table 6.1 defines the REST methods used by the Non-RT RIC to access the resources at the HTTP Server hosted by the Near-RT RIC.

In the opposite direction, the Near-RT RIC is using the POST method to deliver the Feedback policy information.

For example, in the policy creation request, the message refers to the following URI:

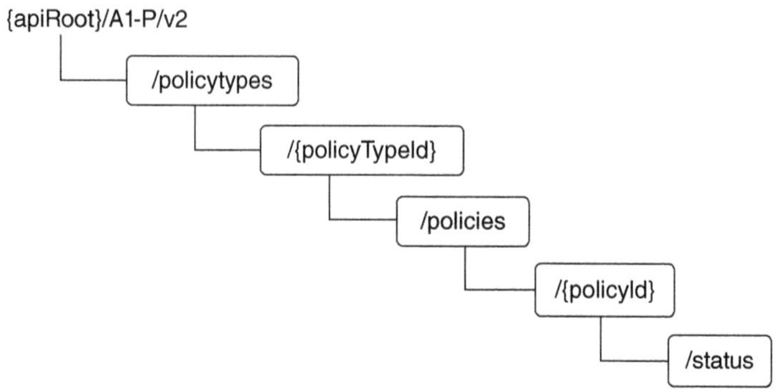

Figure 6.6 Resource URI structure of the A1-P API. (*Source:* Reproduced with permission of O-RAN ALLIANCE e.V.)

Table 6.1 A1 policies and HTTP methods

Operation	**HTTP method**
Create a single policy	PUT
Delete a single policy	DELETE
Query a single policy	GET
Query a single policy type	GET
Query all policy identifiers	GET
Query policy status	GET
Query policy type identifiers	GET
Update a single policy	PUT

(*Source:* Vladimir Yanover)

.../policytypes/{policyTypeId}/policies/{policyId}. The HTTP PUT message body contains the PolicyObject represented with JSON encoding (O-RAN.WG2.A1TD-R003-v07.00 [12], Section 6.4.1). The PolicyObject describes the A1 policy; it contains a "scope" attribute (ScopeIdentifier object) and one or more of policy objective and policy resources statements. Specific policy instance is identified by the policyId included in the URI. While a policy service Producer (Near-RT RIC) can change the enforcement state of the policy instance (ENFORCED vs. NON_ENFORCED), the Producer cannot modify or delete the policy. The Consumer (Non-RT RIC) can subscribe to Policy feedback, in which case it shall provide a callback URI in the query string of the Create policy request. Section 6.3.5 describes the structure of the scope attribute: the scope can be defined as specific individual UE or a group of UEs, etc. More details are provided in following sections.

Other attributes in the PolicyObject represent the following aspects of the policy:

- QoS Objectives
  Specifies requested QoS parameters, defined in 3GPP TS 23.501 [9]. Examples: Guaranteed Flow Bit Rate, Maximum Flow Bit Rate, Packet Delay Budget
- QoE Objectives
  Specifies requested QoE parameters; examples: Mean Opinion Score (MOS), Initial Buffering time, Rebuffering Frequency for video session, etc. See Section 6.3.6
- UE Level Objectives
  Specifies requested UE level QoE parameters with reference to 3GPP definitions in such specifications as 3GPP TS 28.552 [14] and 3GPP TS 38.314 [15]. Examples: average DL/UL UE throughput, DL/UL Packet delay, DL/UL reliability
- Slice SLA Objectives
  Specifies requested SLA targets per NG.116 [16]. Examples: Maximum Number of UEs, Maximum number of PDU sessions, Downlink maximum throughput per UE, Guaranteed downlink throughput quota
- Load balancing Objectives
  Specifies target PRB usage in percentage of the total PRBs in the cell. Covered PRB usage types include mean/peak DL/UL PRBs used for data traffic, either total in the cell or per network slice (per S-NSSAI), where slicing is applied.
- Traffic Steering Preference (TSP) Resources
  Describes restrictions or priorities, which change the traffic scheduling comparatively to the way it would be with default behavior. Examples: set of cells to be avoided or forbidden for certain kind of traffic, set of cells preferred as primary cells for Dual Connectivity
- Slice SLA Resources
  Describes the RAN resources, e.g. cells, available for the system to achieve certain SLA objectives for specific network slice
- Load Balancing Resources
  Describes resources to be used for load balancing; example: set of cells, to which the traffic can be offloaded

6.3.5 Policy Scope

The policy scope ("scope" attribute in the PolicyObject) may contain one or more of the following attributes that identify the objects to which the policy statements are applied; see Sections 6.2 and 6.3 of the "A1 Interface: Type definitions" specification O-RAN.WG2.A1TD-R003-v07.00 [12]:

- Identifier or an individual UE (based on RAN UE Id per 3GPP TS 38.463 [17] and TS 38.473 [18]) or a group of UEs

- QoS Flow identifier
- Network slice identifier
- Cell identifier

6.3.6 Policy Objectives

The policy objectives statement defines the goal for the policy in terms of performance indicators of different types: QoS, QoE, KPI, KQI. The following is the list of the corresponding attributes in the REST realization, with some examples:

- QosObjectives attribute defines QoS Targets per 3GPP TS 23.501 [9]; examples:
  - Guaranteed Flow Bit Rate
  - Maximum Flow Bit Rate
  - Priority Level
  - Packet Delay Budget
- QoeObjectives attribute defines QoE Targets (ITU-T P.1203.3 [19] or proprietary)
  - Mean Opinion Score (MOS) value, e.g. video MOS or proprietary
  - **Initial Buffering**: maximum delay of the start of the video playback from the initiation of the playback by the user
  - **Rebuffering Frequency**: calculated as ratio of the number of stalling events to the length of the media
  - **Stall Ratio**: total duration of the stalling events divided by the total media length
- UeLevelObjectives attribute type defines UE level objectives (3GPP TS 28.552 [14], TS 38.314 [20]), such as
  - Average DL and UL RAN UE throughput
  - DL and UL Packet delay
  - DL and UL Packet loss rate
  - DL and UL reliability: array of
    - Data packet size in bytes
    - The time needed for delivery of a Layer 2/3 SDU to same layer at the other side of the radio link, in ms
    - The probability of successful transmission of a data packet of the specified size within the specified user plane latency
- SliceSlaObjectives attribute type defines Slice SLA targets per GSMA NG.116 [16]:
  - The target for maximum number of connected UEs and maximum number of PDU sessions to be simultaneously served by the network slice
  - Guaranteed data rates in DL and UL to be provided by the network slice
  - Maximum data rates supported by all UEs together in the network slice in DL and UL

  - Maximum data rates supported by the network slice per UE in DL and UL
  - Maximum delay for DL and UL packets in ms
  - Maximum DL and UL PDCP SDU level packet loss rate
  - Minimum DL and UL reliability: array of
    - Data packet size in bytes
    - The time needed for delivery of a Layer 2/3 SDU to same layer at the other side of the radio link, in ms
    - The probability of successful transmission of a data packet of the specified size within the specified user plane latency
  - Maximum DL and UL jitter in ms
  - Priority of the slice in DL and UL
- LbObjectives attribute defines Load balancing objectives (3GPP TS 28.552 [14])
  - The target for PRB usage, in percent
  - Mean and peak DL and UL PRBs used for data traffic
  - Mean and peak DL and UL PRBs used for data traffic per S-NSSAI

For more detailed definitions of policy objectives attributes, see Sections 6.2 and 6.3 of the "A1 Interface: Type Definitions" specification O-RAN.WG2.A1TD-R003-v07.00 [12].

6.3.7 Policy Resource Statement

A statement for policy resources defines the conditions for use of RAN resources under the policy. These conditions can be related to, e.g. cell preferences. For reference, see Section 6.3.4 of O-RAN.WG2.A1TD-R003-v07.00 [12].

The policy resources attributes can be grouped in the following categories:

- Attributes instructing the Near RT RIC how to steer traffic to specific cells in a way different from default
- Attributes indicating the RAN resources (e.g. cells or tracking areas) to be provided to reach the SLA objectives for specific slice
- Attributes used to control traffic offload from a congested cell to indicated candidate cells

One example is the set of attributes associated with Traffic Steering preferences. The TS preferences are encoded as a list (array) of the TspResource objects (O-RAN.WG2.A1TD-R003-v07.00 [12], Section 6.3.4.2). The TspResource object contains the following attributes, which guide the Near-RT RIC in the part of selection of cell resources:

- Attribute "cellIdList" represents the list of Cell IDs; every Cell ID being a composition of PLMN Identifier and Cell Identifier defined in 3GPP

- Attribute "preference" representing the preference of cell usage. Possible values:
  - SHALL guides the Near-RT RIC to select the cell resource unconditionally, even at risk of violation of the connection retainability
  - PREFER guides selection of the cell where the retainability of the connection is not at risk, but not necessarily where the best radio quality can be achieved
  - AVOID is a guidance to avoid selecting the cell resource unless the connection retainability is compromised
  - FORBID is a guidance to not select the cell resource under any conditions
- Boolean attribute "primary" indicates whether the rules are applicable to the selection of primary cell. If the value is FALSE, the rule is considered applicable to secondary cells

Table 6.2 contains a summary guidance for interpretation of the TspResource object.

Another example: the SliceSlaResources object, which indicates the RAN resources (cells or TAs) targeted to specific slice SLA objective:

- List of Cell IDs
  Array of Cell IDs where the slice instance service is supplied. The format of the Cell ID is defined in 3GPP
- List of TAIs
  Array of Tracking Area Identities where the slice instance service is supplied. The format of the TAI is defined in 3GPP

Table 6.2 interpretation of the TspResource object

Value of the attribute "preference"	**Value of the attribute "primary"**	
	true	**false**
SHALL	Only a cell from the cellIdList can be used as primary cell	Only the cells in the cellIdList can be used as secondary cells
PREFER	Preferred to choose the primary cell from the cellIdList. The cellIdList is sorted in the order of descending priority	Preferred to use the cells in the cellIdList as secondary cells. The cellIdList is sorted in the order of descending priority
AVOID	No cell in the cellIdList can be used as primary cell.	No cells in the cellIdList can be used as secondary cells.
FORBID	No cell in the cellIdList can be used as primary cell	No cells in the cellIdList can be used as secondary cells

6.3.8 Example of A1 Policy

For the example, we consider the deployment in Figure 6.7. The target area is covered by two gNBs, each implementing Carrier Aggregation with one 5G Low band cell and one High band cell:

	Low band cells	High band cells
gNB-100	101	102
gNB-200	201	202

The example A1 policy aims differentiation in allocation of the frequency resources between the control plane (RRC) and user plane. To achieve this goal, the policy sets the preferences for selection of Primary and Secondary cells in Carrier Aggregation.

The scope of the policy is defined as a category of UEs associated with a RAN slice subnet with given S-NSSAI. These UEs are provisioned with MBB service.

The guidance for the Near-RT RIC is to assure that for the UEs in the scope:

- Control plane (RRC) is always allocated to low band cells having wider coverage
- User plane MBB data is preferably allocated to high band cells

The solution shown in Table 6.3 defines the A1 policy in terms of the corresponding attributes per O-RAN.WG2.A1TD-R003-v07.00 [12]. The policy instance URI is identified by policyTypeId = "11" and policyId = "1." The policy scope is defined as the set of UEs served under the given network slice with S-NSSAI

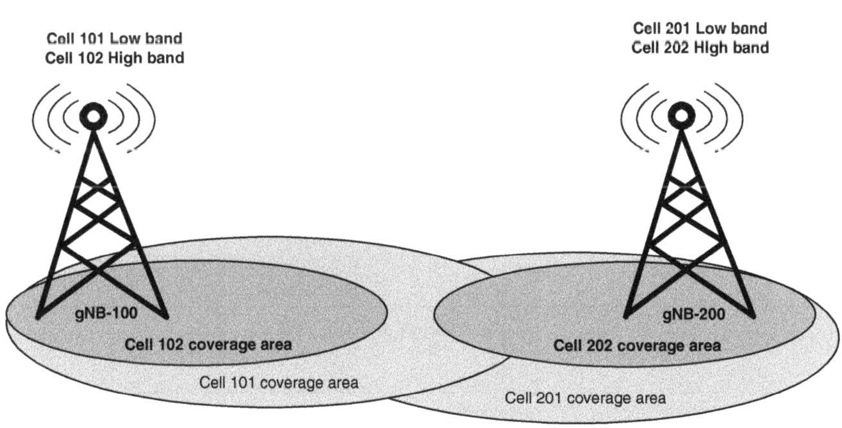

Figure 6.7 Example of A1 policy. (*Source:* Vladimir Yanover.)

Table 6.3 Example of A1 policy

PolicyObject attribute	**Sub-attribute**	**Sub-attribute**	**Value, notes**
URI: .../policytypes/11/policies/1			The policy of the type: policyTypeId = "11" The policy instance ID: policyId = "1"
Scope			
	sliceId		Identifies UEs, which belong to specific slice
		sst	sst = 5 The Slice/Service type as encoded in the S-NSSAI
		sd	sd = 0xDADFCE Slice Differentiator part of S-NSSAI
tspResources			The list of TspResource objects
	tspResource(1)		A TspResource object: Attributes used to schedule traffic on available cells in a way deviated from default behavior
		cellIdList	101, 201 NR Cell IDs for control plane (RRC)
		preference	SHALL only a cell in the cellIdList can be used as primary cell
		primary	True the condition is about primary cells
	tspResource(2)		A TspResource object: Attributes used to schedule traffic on available cells in a way deviated from default behavior
		cellIdList	102, 202, 101, 201 NR Cell IDs in descending order of priority for setting the cells as Secondary
		preference	PREFER Cells in the cellIdList are preferred for use as secondary cells. It is assumed that the cells, which appear earlier in the list, have higher priority
		primary	False The condition is about secondary cells

components: sst = 5 and sd = 0xDADFCE. There are two policy statements, which contain policy resources objects tspResource(1) and tspResource(2). The first statement restricts choice of CA Primary cell to low band cells 101, 201. The second statement sets priority for choice of Secondary cells in the order: 102, 202, 101, 201, so that the high band cells 102, 202 have higher priority than low band cells 101, 201.

6.3.9 A1 Policy Service Operations and Procedures

6.3.9.1 Introduction

Specifications of A1 interface define several A1 Policy-related operations, such as Create policy and Update policy (O-RAN.WG2.A1AP-R003-v04.01 [8], Section 5.2.4). Every such operation is implemented with a HTTP method. Figure 6.8 shows the corresponding service architecture where both A1-P service consumer (Non-RT RIC) and A1-P service producer (Near-RT RIC) contain HTTP client and HTTP server. All operations, except "Notify policy status change," use the HTTP server functionality in the Near-RT RIC. The "Notify policy status change" operation is using HTTP server functionality in the Non-RT RIC. The Near-RT RIC sends policy notifications with POST, using the target URI notificationDestination, received at the step of policy creation.

Figure 6.8 shows mapping of A1 policy operations to HTTP methods. More specifically:

- Create policy and Update policy operations use HTTP method PUT, with HTTP Server located at the Near-RT RIC
- Query policy identifiers, Query policy, and Query policy status operations use HTTP method GET, with HTTP Server located at the Near-RT RIC
- Notify policy status change operation is using HTTP method POST, with HTTP Server located in the Non-RT RIC

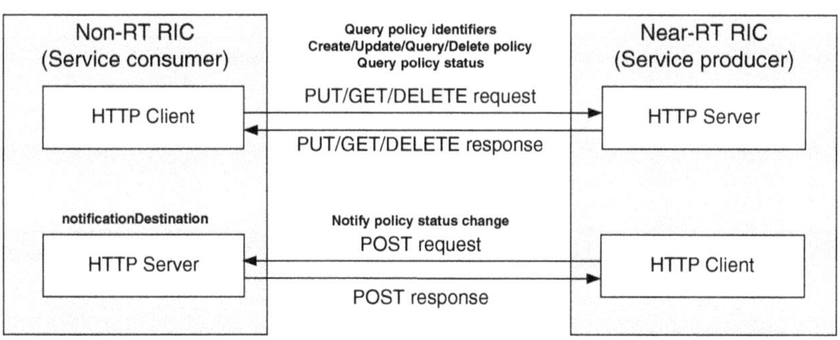

Figure 6.8 HTTP roles in the A1-P service framework. (*Source:* Vladimir Yanover.)

Several policy operations can be combined into a single A1 Policy procedure. For example, the procedure Create multiple policies is implemented as a sequence of several Create policy operations.

Table 6.4 contains the list of the policy operations and the corresponding procedures.

Several following sections describe some of the policy operations and corresponding procedures.

6.3.9.2 Create Policy and Update Policy Operations

This section describes Create policy operation. The Update policy operation works in same way, with the only difference that the policy to be updated, is assumed to exist already. It is explicitly stated in O-RAN.WG2.A1AP-R003-v04.01 [8] that in case the policy does not exist, the request shall be interpreted as Create single policy.

Create policy operation is using the HTTP PUT. The URI (see Section 6.3.2) contains the policyTypeId and the policyId; the message body contains policy description encoded as a PolicyObject.

Figure 6.9 contains the sequence diagram for Create policy operation (successful case).

Table 6.4 Policy operations and supported procedures

Service operation for A1 policies	Procedures based on the operation
Query policy identifiers	Query policy identifiers Query all policy identifiers
Create policy	Create single policy Create multiple policies
Update policy	Update single policy Update multiple policies
Query policy	Query single policy Query multiple policies Query all policies
Delete policy	Delete single policy Delete multiple policies
Query policy status	Query policy status
Notify policy type status change (aka Feedback policy operation)	Feedback policy

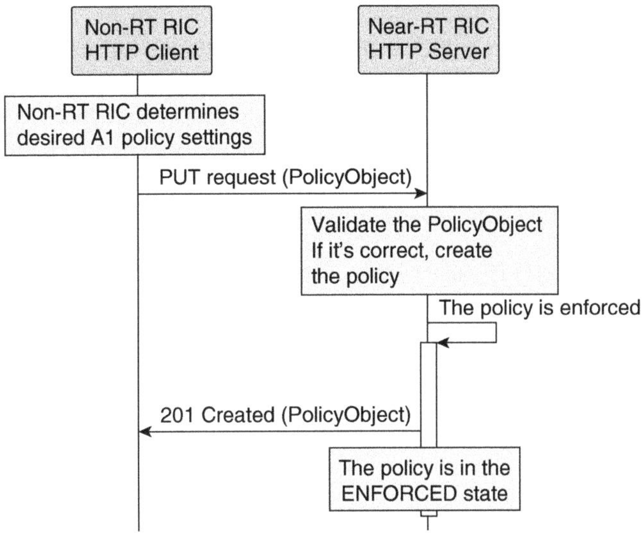

Figure 6.9 Create policy operation, successful case. (*Source:* Vladimir Yanover.)

The operation includes the following steps:

- The Non-RT RIC generates the policyId and sends an HTTP PUT request to the Near-RT RIC. The PUT request contains the URI, under which the policy resource shall be created ".../policytypes/{policyTypeId}/policies/{policyId}." The content of the policy, encoded as a PolicyObject, is included into the message body. The Non-RT RIC may subscribe for policy status updates with notificationDestination included in the PUT request.
- The Near-RT RIC selects proper schemes for validation of the PolicyObject, based on the policyTypeId. If the validation result is positive, the Near-RT RIC creates the policy and sets it to the ENFORCED state.
- The Near-RT RIC returns the HTTP PUT response message with code 201; it contains the URI of the newly created policy and a PolicyObject in the message body.
- The policy remains in the ENFORCED state

In case of failure, the Near-RT RIC returns the corresponding error code.
The Create policy operation can be used in the procedures:

- Create single policy procedure
- Create multiple policies procedure (multiple invocations of Create single policy)

6.3.9.3 Query Policy Operation

The Non-RT RIC can read the policy information from the Near-RT RIC using the Query policy operation. This operation is based on the HTTP GET. The target policy is identified with an URI, which includes the parameters policyTypeId and policyId; the message body is empty. The body of the response message contains the PolicyObject carrying the requested information.

Figure 6.10 describes the Query policy operation with the following steps:

- The Non-RT RIC sends an HTTP GET request message to the Near-RT RIC. The message contains the URI ".../policytypes/{policyTypeId}/policies/{policyId}," which includes policyTypeId and the policyId to identify the target policy; the message body is empty.
- If the policy with the given policyTypeId and policyId exists, the Near-RT RIC generates the response message, carrying the PolicyObject with requested information in the message body
- The Near-RT RIC returns the HTTP GET response message with the code 200 in case of success; the message body contains the PolicyObject, where the requested information is encoded.

In case of failure, the corresponding error code shall be returned.

The following procedures are using the Query policy operation:

- Query single policy
- Query multiple policies
- Query all policies

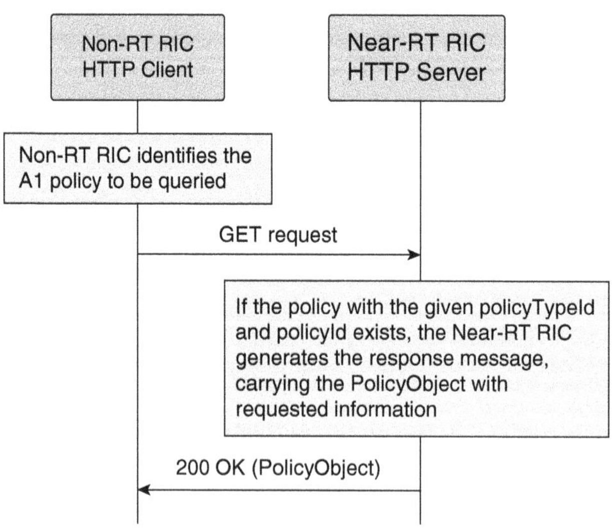

Figure 6.10 Query policy operation, successful case. (*Source:* Vladimir Yanover.)

In case when information on multiple policies is needed, the execution of the procedure includes repeating Query policy operation for all known values of policyId and policyTypeId.

6.3.9.4 Delete Policy Operation

The Non-RT RIC can use the Delete policy operation to delete an A1 policy. The operation is using HTTP DELETE message, which contains the policyTypeId and the policyId in the URI part (Figure 6.11).

The Delete policy operation includes the following steps:

- The Non-RT RIC sends HTTP DELETE request message to the Near-RT RIC, with the URI "…/policytypes/{policyTypeId}/policies/{policyId}" pointing to the targeted policy.
- The Near-RT RIC checks if the policy with given attributes policyTypeId and policyId exists. If yes, it deletes the policy and returns the code 204.

In case of failure, the corresponding error code is returned.

The following procedures are using the Delete policy operation:

- Delete single policy
- Delete multiple policies

The procedure Delete multiple policies is performed by repeating Delete policy operation several times.

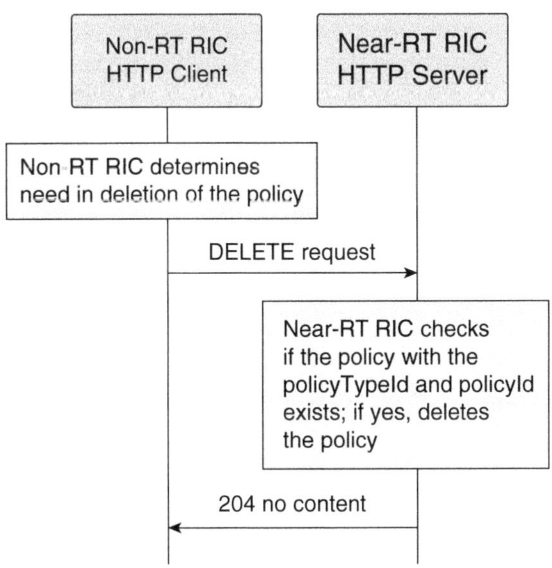

Figure 6.11 Delete policy operation, successful case. (*Source:* Vladimir Yanover.)

6.3.9.5 Query Policy Status Operation

The Query policy status operation is used to read the status of the A1 policy. The status of the policy is represented by the PolicyStatusObject, which contains the following data (O-RAN.WG2.A1TD-R003-v07.00 [12], Section 6.4.2):

- Enforcement status of policy
  It's EnforcementStatusType, which can have one of two values: ENFORCED (the policy is enforced) and NOT_ENFORCED (the policy is not enforced).
- Reason for the latest change of the enforcement status, which can be:
  - The scope (defined in the ScopeIdentifier object) cannot be applied
  - Policy statement(s) cannot be applied
  - Any other reason

The Query operation is using HTTP GET, where the URI includes the parameters policyTypeId and the policyId; the message body is empty. The response returns a PolicyStatusObject.

Figure 6.12 shows the steps in the Query policy status operation:

- The Non-RT RIC sends an HTTP GET request message to the Near-RT RIC. The message contains the URI ".../policytypes/{policyTypeId}/policies/{policyId}/status" identifying the target policy; the message body is empty.
- The Near-RT RIC checks if the policy with given attributes policyTypeId and policyId exists. If yes, it sends the HTTP GET response with the code 200 and the PolicyStatusObject in the body.

The PolicyStatusObject includes information of the current status and the reason why the status changed: enforceStatus (enumerated type; possible values: ENFORCED,

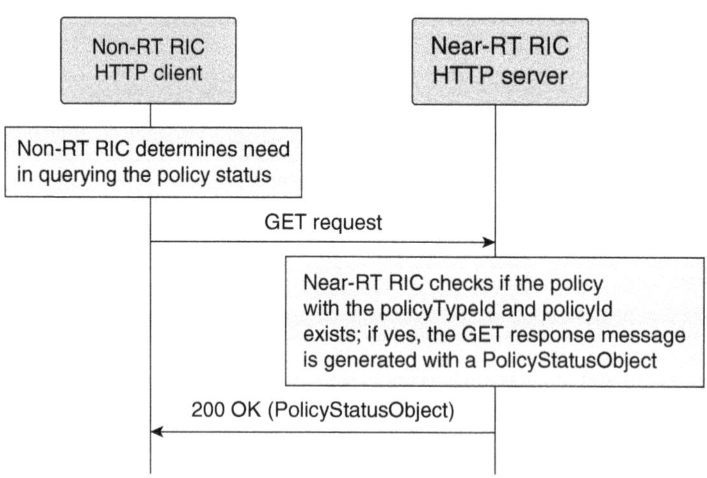

Figure 6.12 Query policy status operation, successful case. (*Source:* Vladimir Yanover.)

NOT_ENFORCED) and enforceReason (enumerated type; possible values: SCOPE_NOT_APPLICABLE, STATEMENT_NOT_APPLICABLE, OTHER_REASON).

In case of failure, the corresponding code is returned.

The following procedures are using the Query policy status operation:

- Query policy status procedure

6.3.9.6 Notify Policy Status Change Operation

In some cases, the Near-RT RIC can unilaterally change the status of the A1 policy, for example, from ENFORCED to NOT_ENFORCED. Then the Non-RT RIC can be notified of this change using Notify policy status change operation. Sometimes this operation is called "feedback policy operation." Prior to receiving notifications, the Non-RT RIC has to subscribe for policy status updates, using Create policy operation or Update policy operation. This operation provides the Near-RT RIC with the URI notifcationDestination, to which the notifications should be delivered. In this operation, the Non-RT RIC plays the role of the HTTP Server, while the Non-RT RIC plays the role of the HTTP Client.

For delivery of a notification, the Near-RT RIC is using the HTTP POST request to the notifcationDestination URI. The notification information is encoded in the PolicyStatusObject carried in the message body; it includes the information of status changes and optionally information about the cause of the change.

The operation includes the following steps (Figure 6.13):

- The Near-RT RIC detects the policy status change and sends an HTTP POST request to the Non-RT RIC. The target URI (notificationDestination) identifies

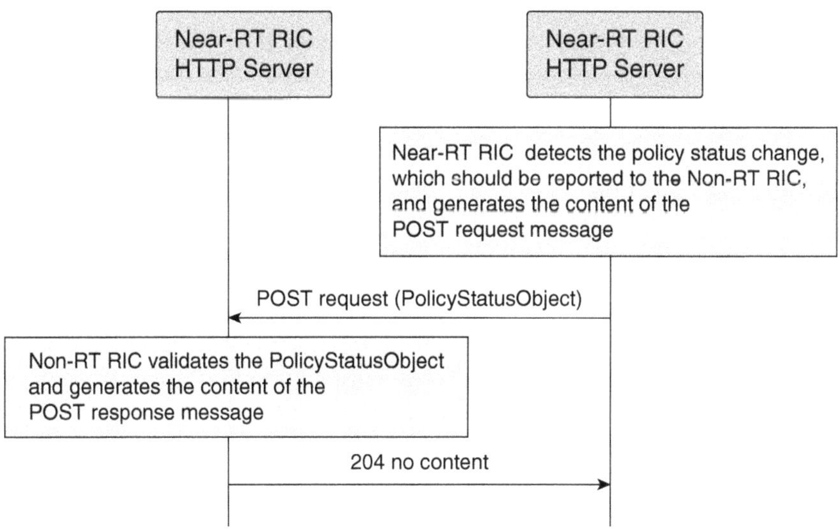

Figure 6.13 Feedback policy operation, successful case. (*Source:* Vladimir Yanover.)

configured earlier URI target for policy status notifications. The message body contains a PolicyStatusObject.

- The Non-RT RIC returns the HTTP POST response with the code 204. The message body is empty.

In case of failure (e.g. validation failure), the Non-RT RIC returns the corresponding code.

Notify policy status change operation is used in the Feedback policy procedure.

6.4 A1 Enrichment Information Service

6.4.1 Introduction

Enrichment Information (EI) can be used by the Non-RT RIC and Near-RT RIC for improvement of RAN control/optimization capabilities (O-RAN.WG2.A1AP-R003-v04.01[8], Section 5.3). The Enrichment Information sources can be internal or external to RAN. Near-RT RIC receives the Enrichment Information from the Non-RT RIC over the A1 interface.

The structure of the A1 Enrichment Information depends on the source; it is defined in the corresponding JSON schemas. See, for example, Table 8.3.3.2-1 in O-RAN.WG2.A1TD-R003-v07.00 [12], which describes attributes related to UE geo-location and velocity. The Near-RT RIC can use A1 interface for discovery of the EI types available at the Non-RT RIC (EiTypeIds).

The A1 Application Protocol specification O-RAN.WG2.A1AP-R003-v04.01 [8] contains definitions for A1-EI Service, enabling communication between the Near-RT RIC, considered an A1-EI Consumer and the Non-RT RIC considered an A1-EI Producer. For communication, HTTP protocol is used, with same HTTP Server and HTTP Client roles, as for the policy management service A1-P; see Figure 6.8.

6.4.2 EI Job Creation

For generation and delivery of the EI information, the Near-RT RIC (A1-EI Consumer) requests creation of an EI job in the Non-RT RIC. The request contains the job description encoded as a JSON object EiJobObject; see Section 8.4.2 in O-RAN.WG2.A1TD-R003-v07.00 [12]. The structure of the EiJobObject includes identifier of the type of the Enhancement Information, URI for delivery of the EI Job results, and the jobDefinition object.

For example, in case of UE geo-location and velocity information (EI data of the type UeGeoandVelEIDescription), the job definition object contains the attributes:

- gadShape

Geographical Area Description (GAD) shapes used for UE geolocation information: Ellipsoid point, Polygon, Ellipsoid arc, ... etc.

- (Optional) velocityDesc

The UE velocity description type; possible values correspond to: "Horizontal velocity," "Horizontal and vertical velocity," ... etc.

The job definition object also contains such parameters as the interval of periodic measurement and the interval of periodic reporting.

Before requesting creation of the EI Job, the A1-EI Consumer may query the A1-EI Producer about available EI types (Query EI type operation). Response to such query contains EiTypeObject, which includes JSON schemas for validation of EI Job definition, EI Job status, EI Job result, and EI Job constraints. See Section 8.4.1 in O-RAN.WG2.A1TD-R003-v07.00 [12] for more details

6.4.3 A1-EI Resource Identifiers and Operations

In the A1-EI HTTP operations, the Non-RT RIC and Near-RT RIC use URIs, which include the following identifiers: eiTypeId (EI-type identifier string), and eiJobId (EI job identifier string)

Access to Enrichment Information includes the following steps:

- Query EI type identifiers operation
  The Near-RT RIC sends a query to the Non-RT RIC to receive the list of EI types supported by the A1-EI Producer. The response is encoded as an array of EI-type IDs
- Create EI job operation
  After the Near-RT RIC identified a need in receiving certain kind of EI, it requests creation of an EI job at the Non-RT RIC, using PUT operation carrying EiJobObject, which contains attributes of the requested job. The EiJobObject contains also the URI jobResultUri, to which the job results should be sent. If the request is granted, the response with code 201 carries description of the created job.
- Deliver EI job result operation
  After some EI job results are ready, the Non-RT RIC sends this information using POST to the URI provided by the Near-RT RIC at the step of EI job creation. The response message body carries EiJobResultObject, which contains the outcome of the job. The A1-EI responds with confirmation 204 "No Content."

References

1 O-RAN.WG1.OAD-R003-v11.00. O-RAN Architecture Description

2 O-RAN.WG2.Non-RT-RIC-ARCH-R003-v05.00. Non-RT RIC Architecture

3 Vodafone and NTT DOCOMO. SMO Deployment Challenges in an Open RAN Environment. Joint whitepaper. https://www.vodafone.com/sites/default/files/2023-10/vfe-docomo-smo-whitepaper.pdf (accessed 28 October 2023)
4 O-RAN.WG2.R1GAP-v07.00. O-RAN R1 interface: General Aspects and Principles
5 O-RAN.WG2.R1UCR-R003-v06.00. O-RAN R1 interface: Use Cases and Requirements
6 O-RAN.WG2.Use-Case-Requirements-v09.00. Non-RT RIC & A1 Interface: Use Cases and Requirements
7 O-RAN.WG2.A1GAP-R003-v03.02. A1 interface: General Aspects and Principles
8 O-RAN.WG2.A1AP-R003-v04.01. A1 interface: Application Protocol
9 3GPP TS 23.501. System architecture for the 5G System (5GS)
10 3GPP TS 29.501 5G System; Principles and Guidelines for Services Definition; Stage 3
11 3GPP TS 32.158 Design rules for representational State Transfer (REST) Solution Sets (SS)
12 O-RAN.WG2.A1TD-v07.00. A1 interface: Type Definitions
13 O-RAN.WG2.AIML-v01.03. O-RAN AI/ML workflow description and requirements
14 3GPP TS 28.552. 5G performance measurements
15 3GPP TS 38.314 Layer 2 Measurements
16 GSMA NG.116 Generic Network Slice Template Version 7.0 17 June 2022
17 3GPP TS 38.463. NG-RAN; E1 Application Protocol (E1AP)
18 3GPP TS 38.473. NG-RAN; F1 Application Protocol (F1AP)
19 ITU-T P.1203.3. Parametric bitstream-based quality assessment of progressive download and adaptive audiovisual streaming services over reliable transport
20 3GPP TS 36.401. E-UTRAN; Architecture Description

7

Real-Time RAN Programming

CONTENTS

Open RAN: Technology and Ecosystem, First Edition. Vladimir Yanover.

Acronyms and Abbreviations

API	Application Programming Interface
CHO	Conditional Hanover
DAPS	Dual Active Protocol Stack (DAPS)
DRB	Data Radio Bearer
E2AP	E2 Application Protocol
E2SM	E2 Service Model
gNB-CU	gNB – Central Unit
gNB-CU-CP	gNB – Central Unit – Control Plane
gNB-CU-UP	gNB – Central Unit – User Plane
gNB-DU	gNB – Distributed Unit
HO	Handover
IE	Information Element
KPM	Key Performance Measurements
Near-RT RIC	Near-real-time Radio Intelligent Controller
Non-RT RIC	Non-Real-Time Radio Intelligent Controller
MN	Master Node role in the Dual Connectivity
NR	New Radio
O-CU	O-RAN Central Unit
O-CU-CP	O-RAN Central Unit – Control Plane
O-CU-UP	O-RAN Central Unit – User Plane
O-DU	O-RAN Distributed Unit
O-RU	O-RAN Radio Unit
QoS	Quality Of Service
RAN	Radio Access Network
RAN Function	A specific function in an E2 Node; examples include interface functions, RRM, DRB control, CONNECTED state mobility control, and admission control

RBC Radio Bearer Control
rApp Third-party application hosted by the Non-RT RIC
RRC Radio Resources Control
SDK Software Development Kit
SN Secondary Node role in the Dual Connectivity
xApp Third-party application hosted by the Near-RT RIC

7.1 Introduction

In this chapter we discuss the Near-Real-Time Radio Intelligent Controller ("Near-RT RIC"), which is a new element of the RAN architecture defined by the O-RAN ALLIANCE. The following aspects are addressed:

- Near-RT RIC in the O-RAN architecture
- Near-RT RIC Interfaces: E2, A1, O1, and Y1
- Internal Near-RT RIC architecture; hosting of third-party application
- Implementation of RAN control and optimization functionalities; using Near-RT RIC for near-real-time and real-time RAN programming.

7.2 Near-RT RIC in the O-RAN Architecture

Figure 7.1 shows Near-RT RIC in the Logical Architecture of O-RAN (O-RAN.WG1. OAD-R003-v11.00 [1]). In this architecture Near-RT RIC is defined as a controller, hosting some part of RAN intelligence. It is connected to the remaining part of RAN (O-gNB or O-eNB) via the E2 interface. In the case the O-gNB is split according to 3GPP specifications TS 38.410 [2], the corresponding RAN components are named O-CU with possible further split into O-CU-CP and O-CU-UP. The O-DU, which corresponds to defined in 3GPP gNodeB-DU, can be split into O-DU and O-RU. The prefix "O-" signifies that some part of intelligence in the corresponding 3GPP-defined RAN components gNB-CU-CP, gNB-CU-UP, gNB-DU is separated and moved to the Near-RT RIC. The Near-RT RIC has some level of control of the RAN components O-CU (O-CU-CP, O-CU-UP), O-DU via the E2 interface, which is the key enabler of near-real-time RAN programmability. The components connected to the Near-RT RIC via E2 interface are called "E2 Nodes."

The E2 interface capabilities enable near-real-time RAN automation and optimization:

- The Near-RT RIC can collect near-real-time RAN data from E2 Nodes; UE level and cell level information is supported
- The Near-RT RIC can receive near-real-time notifications of events in the E2 Nodes

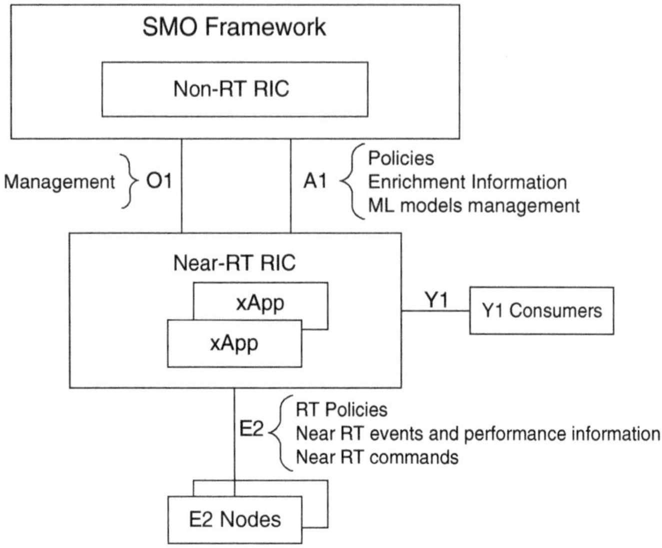

Figure 7.1 Near-RT RIC in O-RAN Architecture. (*Source:* Vladimir Yanover.)

- The Near-RT RIC can perform near-real-time control and optimization actions toward E2 Nodes
- The Near-RT RIC can set policies at the E2 Nodes that define their reaction in real time to specific internal events

The A1 interface allows non-real-time control of the Near-RT RIC, including control of RAN automation and optimization functionalities residing in the Near-RT RIC. The Non-Real-Time Radio Intelligent Controller ("Non-RT RIC") is using A1 for setting non-real-time policies ("A1 Policies"), for delivery of enrichment information to the Near-RT RIC and for AI/ML framework operations; the details can be found in Chapters 6 and 11.

The O1 interface enables network management of Near-RT RIC.

The Y1 interface is used for exposure of the RAN analytics information generated in the Near-RT RIC.

Abovementioned interfaces and capabilities provide support for use cases described in O-RAN.WG3.UCR-R003-v05.00 [3].

Different instances of the E2 interface may connect the Near-RT RIC to E2 Nodes of different kinds, such as an integrated 5G O-RAN node O-gNB, or one of O-RAN components O-CU-CP, O-CU-UP, O-DU. For all these variants, O-RAN specifications O-RAN.WG3.E2AP-R003-v05.00 [4] and O-RAN.WG3.E2GAP-R003-v05.00 [5] define a common set of E2 operations. Additionally, for every kind of E2 Node function, the E2 Service Model ("E2SM") captures specificity of the function. Figure 7.1 also shows third-party applications ("xApps") running on the Near-RT

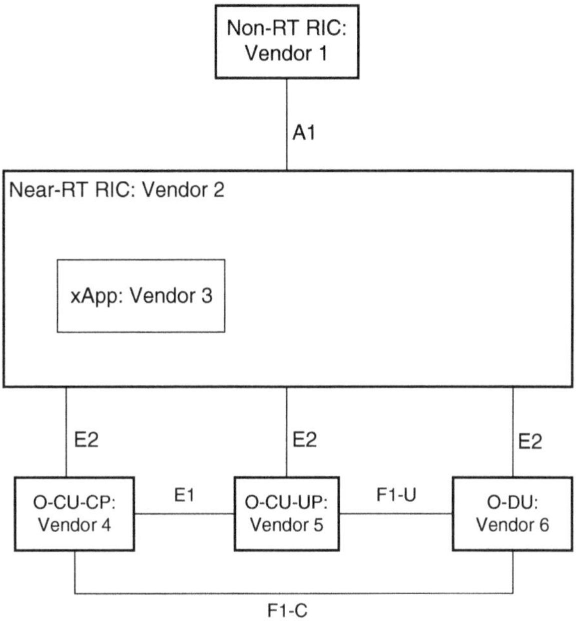

Figure 7.2 Multi-vendor RAN. (*Source:* Vladimir Yanover.)

RIC platform. The Near-RT RIC platform provides xApps with access to E2 and to other Near-RT RIC interfaces.

With the set of standardized interfaces, the O-RAN architecture allows multi-vendor RAN implementation, where different O-RAN Network Functions, including Near-RT RIC, can be manufactured by different vendors. Figure 7.2 shows an (extreme) example of multi-vendor RAN.

7.3 E2 General Aspects and Principles

The O-RAN specification O-RAN.WG3.E2GAP-R003-v05.00 [5] outlines E2 general aspects and principles.

E2 is a logical interface between the Near-RT RIC and a RAN component ("E2 Node"). The E2 Node contains one or more of the Network Functions: O-CU, O-CU-CP, O-CU-UP, and O-DU.

While every E2 Node can be connected to only one Near-RT RIC, multiple E2 Nodes can be connected to same Near-RT RIC. An E2 Node may contain multiple network functions, supporting E2 interface, so a single E2 Node may expose multiple instances of the E2 interface, with same or different E2 Service Models (E2SMs). A network node exposing E2 interface(s) may include network functions, which do not expose the E2 interface. The E2 Setup and RIC Service

Update procedures (O-RAN.WG3.E2GAP-R003-v05.00 [5]) allow Near-RT RIC to discover the RIC services supported by specific E2 Node.

The functions of the E2 protocol can be grouped into two categories: RIC Services and RIC support functions (O-RAN.WG3.E2GAP-R003-v05.00 [5], Section 5.3).

RIC Services

RIC Services category includes the functions involved in implementation of RIC services REPORT, INSERT, CONTROL, POLICY, and QUERY, provided by the E2 Node to the Near-RT RIC. They are introduced in Section 5.3.2 of the specification O-RAN.WG3.E2GAP-R003-v05.00 [5] and described in the O-RAN.WG3.E2AP-R003-v05.00 [4].

In line with modern "service-based" approach, O-RAN RIC services specifications separate the definition of protocol operations from the information model.

In the part of operations, the RIC services are based on the RIC functional procedures defined in O-RAN.WG3.E2AP-R003-v05.00 [4]. The functional procedures related to subscription are RIC Subscription, RIC Subscription Modification, RIC Subscription Delete, RIC Subscription Delete Required, RIC Subscription Modification Required. Common procedures: RIC Control, RIC Indication, and RIC Query. More details on the RIC functional procedures can be found in Section 7.4.

The information models are called "E2 Service Models" ("E2SMs"). For information on the E2 Service Models see Sections 7.5 and 7.6 and specifications O-RAN.WG3.E2SM-CCC-R003-v03.00 [6], O-RAN.WG3.E2SM-KPM-R003-v04.00 [7], O-RAN.WG3.E2SM-R003-v05.00 [8], and O-RAN.WG3.E2SM-RC-R003-v05.00 [9].

The E2SM construction makes the format of E2 messages and Information Elements dependent on selection of several IDs. One of them is the type of the E2 Service, e.g. REPORT. Another is Service Style Type (sometimes called simply "Style"). For example, RIC INDICATION message can be used in the services REPORT (with Styles 1–4) and INSERT (with Styles 1–7). The format of the message depends on the type of the E2 Service, on the selected Style Type, and on some additional parameters. Details of format selection can be found in Section 7.4.3.

RIC support functions

RIC support functions category contains the following.

- **Interface Management procedures**: E2 Setup, E2 Removal, E2 Node Configuration Update, E2 Reset, Reporting of General Error Situations
- **RAN Function service procedures**:
  - RIC Service Update procedure, which can be initiated by the E2 Node or by the Near-RT RIC
  - RIC Service Query procedure

More details on RIC functional procedures can be found in O-RAN.WG3.E2AP-R003-v05.00 [4], Chapter 8.

7.4 RIC Functional Procedures

7.4.1 Introduction

RIC functional procedures are building blocks for definition of the RIC Services. The RIC functional procedures (aka E2AP Elementary Procedures) are defined in O-RAN.WG3.E2AP-R003-v05.00 [4] in terms of the E2 Application Protocol (E2AP). Tables 7.1 and 7.2 provide complete list of Elementary Procedures. Split into Class 1 and Class 2, simplifies description of what the E2 Node is required to do in the case

Table 7.1 Class 1 elementary procedures

Elementary procedure	Initiated by	Description
RIC Subscription	Near-RT RIC	Sets in the E2 Node one or more conditions, for every of which specific E2 Node actions are defined to be triggered once the condition is met
RIC Subscription Delete	Near-RT RIC	Deletion of the subscription
RIC Subscription Modification	Near-RT RIC	Modification of the subscription
RIC Subscription Modification Required	E2 Node	Request for modification of the subscription
RIC Control	Near-RT RIC	Command to the E2 Node, which modifies certain internal configuration parameters or changes the state of the E2 Node or triggers execution of certain action, e.g., handover
RIC Query	Near-RT RIC	Request for RAN- and/or UE-related information from E2 Node
E2 Setup	E2 Node	Establishment of the signaling connection between E2 Node and Near-RT RIC
RIC Service Update	E2 Node	Updates the Near-RT RIC on changes in the list of the E2 services supported by the E2 Node
E2 Node Configuration Update	E2 Node	Update application level configuration data, e.g. TNL association information
E2 Connection Update	Near-RT RIC	Update the TNL information for the E2 connection between the E2 Node and the Near-RT RIC
Reset	Near-RT RIC or E2 Node	Resets the E2 interface, usually in the event of an abnormal failure
E2 Removal	Near-RT RIC or E2 Node	Removal of the E2 interface connection

Table 7.2 Class 2 elementary procedures

Elementary procedure	Initiated by	Description
RIC Indication	E2 Node	Real-time indication of the event ("Event Trigger") which occurs in the E2 Node; the indication is done as transfer of a message, carrying information of the event and possibly additional information. The Event Trigger and the content of the message are defined in the corresponding subscription
RIC Service Update[a]	Near-RT RIC	Update the Near-RT RIC on the set of RAN Functions exposed by the E2 Node via the E2 interface
RIC Subscription Delete Required	E2 Node	E2 Node requests deletion of the existing subscriptions created by request from the Near-RT RIC
Error Indication	E2 Node or Near-RT RIC	Report on the detected errors

[a] Alternative name "RIC Service Query" is used in Table 8.1-2 in the specification O-RAN.WG3.E2AP-R003-v05.00 [5]

when the requested functionality is not supported or the request cannot be served (O-RAN.WG3.E2AP-R003-v05.00 [4], Section 4.1). In such scenario, for Class 1 procedures, the receiving node is required to report unsuccessful outcome of the procedure. There is no such requirement for Class 2 procedures.

It is important to note that some E2AP Information Elements, used in the Elementary Procedures, appear in the E2AP specification O-RAN.WG3.E2AP-R003-v05.00 [4] as OCTET STRING with note "Defined in RAN Function specific E2 Service Model." The format of such messages is RAN Function specific or E2 Service specific. Important examples are:

- RIC Event Trigger Definition and RIC Action Definition used in Subscription procedure
- RIC Control Header and RIC Control Message used in Control procedure
- RIC Indication Header and RIC Indication Message used in Indication procedure
- RIC Query header and RIC Query Definition used in Query procedure

The format of such IEs may differ in different E2SMs and even when used in different services within same E2SM. For example, the RIC Action Definition Information Element in the RIC SUBSCRIPTION REQUEST message in the E2SM-RC is defined differently for the REPORT, POLICY, and INSERT services; see O-RAN.WG3.E2SM-RC-R003-v05.00 [9], Section 9.2.1.2.

Next sections contain detail description of several important Elementary Procedures.

7.4.2 RIC Subscription Procedure

7.4.2.1 Introduction

The Subscription procedure is an essential element of the REPORT, INSERT, and POLICY Services. It is used for programming the E2 Node's reaction to certain events ("Event Triggers"). This procedure creates RIC Subscriptions at the E2 Node. Every subscription defines an Event Trigger linked to a list of RIC Service Actions. For example, in the case of REPORT service, the subscription instructs the E2 Node to generate and send specific network information, such as RAN measurements or the state of specific RAN functions. Usually, for the REPORT Service, the Event Trigger is periodic. In the case of INSERT Service, the subscription instructs the E2 Node to send the RIC INDICATION message to the Near-RT RIC immediately after detection of a RAN event, such as reception of an RRC message [10]. The subscription defines the RAN event, the content of the RIC INDICATION message, and more.

It is important to note that the O-RAN specifications implicitly assume that for any kind of RAN event in the E2 Node, there is an internally defined procedure for handling the events of this kind; it is called "associated procedure"; sometimes (historical) term "call process" is used.

For the INSERT Service, the associated procedure is executed every time after the Event Trigger is detected, unless different instructions are set in the subscription. In the case of INSERT, the associated procedure is suspended until either a response with instructions is received from the Near-RT RIC, or the Time to Wait timer is expired. Some details can be found in Section 7.5.5.

7.4.2.2 RIC Subscription Attributes

The subscription defines an Event Trigger and a sequence of actions. The subscription attributes are transferred from the Near-RT RIC to the E2 Node in the RIC SUBSCRIPTION REQUEST message. The message contains the RAN Function ID and RIC Request ID; together they provide unique identity for the subscription request within the E2 Node. The message also contains start and stop time for the subscription. In addition, the message contains information of the requested subscription, RIC Subscription Details IE (O-RAN.WG3.E2AP-R003-v05.00 [4], Section 9.1.1.1), which includes:

- **RIC Event Trigger Definition**: In the E2AP protocol, it is specified as an OCTET STRING, so exact definition is RAN Function specific and belongs to the E2 Service Model
- **Sequence of Actions, a list, every element of which contains**:
  - RIC Action ID: Unique ID within the given RIC SUBSCRIPTION REQUEST
  - RIC Action Type (with possible values Insert, Report, and Policy): Identifies the type of the E2 Service where the Subscription procedure is used; one of: INSERT, REPORT, and POLICY
  - (Optional) RIC Action Definition. In the E2AP protocol, it is specified as an OCTET STRING; exact definition is RAN Function specific and belongs to

the E2 Service Model. Notably, the service model E2SM-RC defines the format of RIC Action Definition in different way for the REPORT, POLICY, and INSERT services; see O-RAN.WG3.E2SM-RC-R003-v05.00 [9], Section 9.2.1.2.
- (Optional; for the RIC Action Type = Insert only) RIC Subsequent Action. This information element determines the action to be performed after the RIC Action is completed; possible values: Continue, Halt. Details can be found in Section 7.4.3 on the RIC Indication procedure.
- (Optional) RIC Action Execution Order: in the case, when two RIC Actions have different values of the RIC Action Execution Order, the one with smaller value is executed earlier

Figure 7.3 shows an example flow with a RIC Subscription procedure and subsequent RIC Indication procedure. The Near-RT RIC starts the procedure by sending the RIC SUBSCRIPTION REQUEST message. In the successful case, the E2 Node configures the requested event triggers, stores information of the requested actions, and sends back to the Near-RT RIC the RIC SUBSCRIPTION RESPONSE message, where admitted and not admitted RIC Actions can be listed.

Alternatively, in the case when the requested subscription cannot be established, the E2 Node sends a RIC SUBSCRIPTION FAILURE message.

After the subscription is established, the E2 Node starts monitoring the conditions included into the RIC Event Trigger Definitions. When an Event Trigger is detected, the E2 Node does the following:

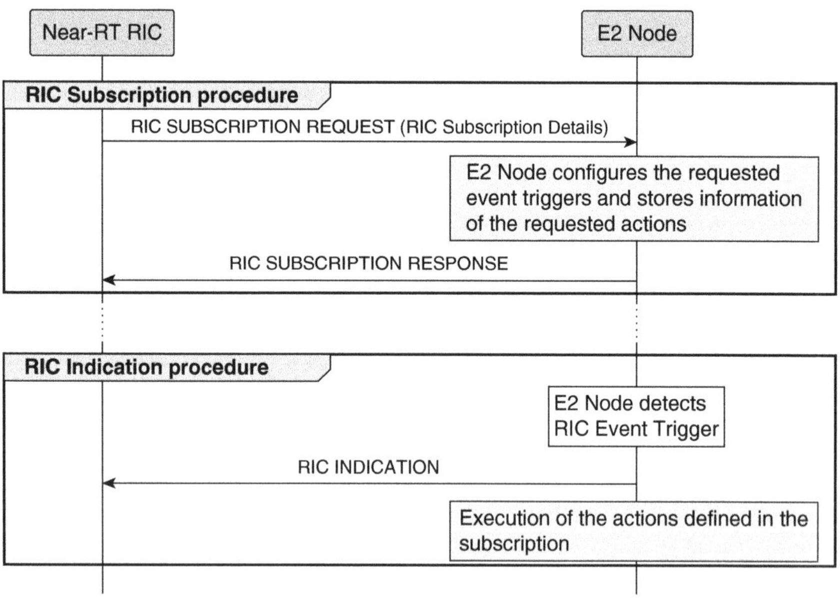

Figure 7.3 RIC Subscription and RIC Indication procedures. (*Source:* Vladimir Yanover.)

- Stores current state of the associated procedure
- Sends RIC INDICATION message with RIC Call Process ID, which identifies the instance of the associated procedure. For some E2 Services, the associated procedure can be suspended

7.4.3 RIC Indication Procedure

The Indication procedure is an essential element of the REPORT and INSERT services. It is used for sending a real-time indication ("RIC INDICATION" message) to the Near-RT RIC immediately after the Event Trigger is detected. In the case of REPORT, the Indication procedure is used to generate and send the network information defined in the subscription, such as RAN measurements and counters; usually, the Event Trigger is periodic. For the INSERT service, the RIC INDICATION message may contain real-time information related to the Event Trigger. The Near-RT RIC can use this information to react to the event in near real time. For example, if the procedure was triggered by reception of an RRC MeasurementReport message [10], RIC INDICATION message may carry information of the sender identity and the reported measurement results. Such use of RIC Indication procedure enables the Near-RT RIC to intervene in O-CU handover decisions.

The bottom part of Figure 7.3 shows the flow of the RIC Indication procedure. E2 Node executes the procedure by sending the RIC INDICATION message (the word "INDICATION" is fully capitalized to differentiate from "RIC Indication Message IE"). In the E2SM-RC, the RIC INDICATION message contains several Information Elements:

- **RIC Indication Type IE**
  Identifies the type of E2 service where the RIC INDICATION message is used: Insert or Report
- **RAN Function ID IE**
  Identifies the RAN Function, which generated the RIC INDICATION message
- **RIC Request ID IE**
  The ID, assigned by the Near-RT RIC as a part of the RIC Subscription procedure for association of the RIC INDICATION message with specific subscription
- **RIC Action ID IE**
  Identifier of the specific Action within specific RIC Subscription Request, which caused the RIC INDICATION message
- **RIC Indication Header IE**
  Contains RIC Insert Style Type; may contain other parameters
- **RIC Indication Message IE**
  Names RAN parameters, for which the E2 Node is requesting the Near-RT RIC to set the values, via the E2 CONTROL action, in response to the RIC INDICATION message.

7.4.4 RIC Control Procedure

The RIC Control procedure is used by the Near-RT RIC to modify behavior of the E2 Node. In particular, it can be the behavior of the E2 Node in handling its internal event; the Near-RT RIC's intervention can be done in near-real time after reception of RIC INDICATION message.

The Near-RT RIC starts the procedure by sending to the E2 Node a RIC CONTROL REQUEST message (Figure 7.4), which contains several Information Elements:

- **RIC Request ID IE**
  Assigned by the Near-RT RIC
- **RAN Function ID IE**
  Identifies the RAN Function to which the message is addressed
- **RIC Control Header IE**
  Contains such parameters as RIC Control Style Type, RIC Control decision (accept or reject the request received in the RIC INDICATION message within the E2 INSERT Service), and Control Action ID (identifies the action requested from the E2 Node).
- **RIC Control Message IE**
  Specifies Control actions, which Near-RT RIC requests to execute; in particular, it can be a request for modification of internal parameters in the target E2 Node

The message may also contain request for acknowledgement: RIC Control Ack Request IE.

The Control procedure can be coupled with immediately preceding Indication procedure. In such scenario, the Insert procedure informs the Near-RT RIC of the corresponding Event Trigger, which occurred in the E2 Node. After processing the received RIC INDICATION message and making certain determination, the

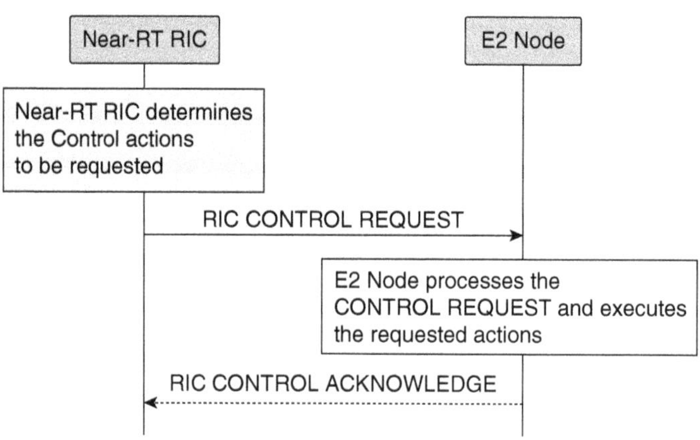

Figure 7.4 RIC Control procedure. (*Source:* Vladimir Yanover.)

Near-RT RIC is using the Control procedure to transfer the corresponding information to the E2 Node. For example, the Near-RT RIC may determine specific handover target and transfer the identification of the target to the E2 Node (O-CU). In such scenario, the RIC CONTROL REQUEST message will contain RIC Call Process ID IE, received in the RIC INDICATION as identification of the associated procedure. The RIC CONTROL REQUEST message will also contain instruction on how to handle the triggering event.

After the E2 Node receives the RIC CONTROL REQUEST message, it shall process the received information and execute on it. In the case the message responds to prior RIC INDICATION message, the received information normally resumes the suspended associated procedure and may significantly influence continuation of the associated procedure. For example, in the case the Indication was triggered by reception of an RRC MeasurementReport message from a UE [10], the RIC CONTROL REQUEST may contain information on selection of specific handover target cell. Then the E2 Node (O-CU) initiates handover to the specified cell.

The Near-RT RIC can also generate RIC CONTROL REQUEST message without any relation to prior RIC INDICATION message. For example, the RIC CONTROL REQUEST message may include indication of CONTROL Service Style 1, which means it targets Radio Bearer control (O-RAN.WG3.E2SM-RC-R003-v05.00 [9], Section 7.6.2). Then the requested action can be one of the following:

- Data Radio Bearer (DRB) QoS Modification, which is used to modify DRB level QoS parameters
- QoS Flow (re)mapping, which can be used to modify the mapping between the QoS flows and the DRBs.
- Logical Channel (re)configuration
- Action related to Radio Bearer Admission Control or DRB Split Ratio Control or PDCP duplication control

If the RIC CONTROL REQUEST contains request for acknowledgement, the E2 Node responds with RIC CONTROL ACKNOWLEDGE message.

7.4.5 RIC Query Procedure

The Near-RT RIC begins the Query procedure with the RIC QUERY REQUEST message. The message contains identification of the target RAN function and request for information. If the requested information is available, the E2 Node responds to the Near-RT RIC by sending the RIC QUERY RESPONSE message, which contains the requested information.

In the E2SM-RC, two Service styles are defined:

- **E2 Node Information Query**
  Examples of E2 Node information: Cell Context Information, Neighbor Relation Table.

- **UE Information Query**
  Example of UE Information: the Serving Cell ID.

7.5 E2 Services

7.5.1 Introduction

The RIC Services (also referred to as E2 Services) are defined in O-RAN.WG3.E2GAP-R003-v05.00 [5]. The RIC Services are based on the RIC functional procedures described in Section 7.4. For all RIC Services, the E2 Nodes are service providers, and the Near-RT RIC is the service consumer.

Functionality of every E2 Service essentially depends on the selected E2 Service Model. In this section, examples of service operations are based on the E2SM-RC; for the REPORT service also on E2SM-KPM.

Table 7.3 shows mapping between E2 Services and RIC Functional Procedures. For example, implementation of the REPORT service includes use of two RIC Functional Procedures: RIC Subscription and RIC Indication.

7.5.2 E2 Services Modeling Scheme

7.5.2.1 Selection of Formats and RAN Parameters

In an E2SM, selection of the format of E2AP messages/IEs depends on selection of several IDs, such as E2 Service, Service Style, and other parameters. The goal of this section is to map selection of such IDs to the section(s) of the E2AP specification where the corresponding message formats are defined. As The Event Trigger definitions do not depend on the Service and Service Style selection, they are considered in separate Section 7.5.2.2.

Table 7.3 Mapping between E2 Services and RIC functional procedures

E2 Service	RIC functional procedures
REPORT	RIC Subscription RIC Indication
INSERT	RIC Subscription RIC Indication
CONTROL	RIC Control
POLICY	RIC Subscription RIC Indication

Figure 7.5 shows the scheme of how required message format and the list of RAN Parameters can be selected; all references in the figure point to the E2SM-RC specification O-RAN.WG3.E2SM-RC-R003-v05.00 [9]. The following IEs are covered in Figure 7.5:

- RIC Action Definition IE
- RIC Indication Header IE
- RIC Indication Message IE
- RIC Call Process ID IE
- RIC Control Header IE
- RIC Control Message IE
- RIC Control Outcome IE

The following are steps in Figure 7.5:

- **Step 1**: Selection of the E2 Service Model (E2SM)
  In all examples, the E2SM-RC is used.
- **Step 2**: Selection of E2 Service
  One of REPORT, INSERT, CONTROL, POLICY.
- **Step 3**: Selection of Service Style Type

For every RIC Service, multiple Style Types (sometimes called simply "Styles") are defined. These definitions can be found in O-RAN.WG3.E2SM-RC-R003-v05.00 [9]. For the REPORT service it's Section 7.4.1 in the E2SM-RC specification, for INSERT service it's Section 7.5.1, for CONTROL service it's Section 7.6.1, and for POLICY service it's Section 7.7.1. Examples of INSERT Service Styles:

- (**Style 1**) Radio Bearer Control Request Style is used for requesting modification of the parameters, related to DRB control, such as 5QI, Packet Delay Budget, and Maximum/Guaranteed Flow Bit Rate Downlink/Uplink
- (**Style 2**) Radio Resource Allocation Control Request is used for requesting modification of the Radio Resource Allocation parameters; examples: resourceAllocation, sps-Config configuration parameters, and cqi-Table, defined in 3GPP TS 38.331 [10].
- (**Style 3**) Connected Mode Mobility Control Request is used for triggering of CONNECTED state mobility procedures, such as Handover, Conditional Handover (CHO), and Dual Active Protocol Stack (DAPS). The mobility procedure for specific UE identifies specific handover target cell(s)

Other Styles are defined for Radio Access Control Request, Dual Connectivity Control Request, Carrier Aggregation Control Request, Idle Mode Mobility Control Request (Section 7.5.5.3).

- **Step 4** (RAN parameters only): selection of Indication ID (INSERT Service) or Control Action ID (CONTROL Service)

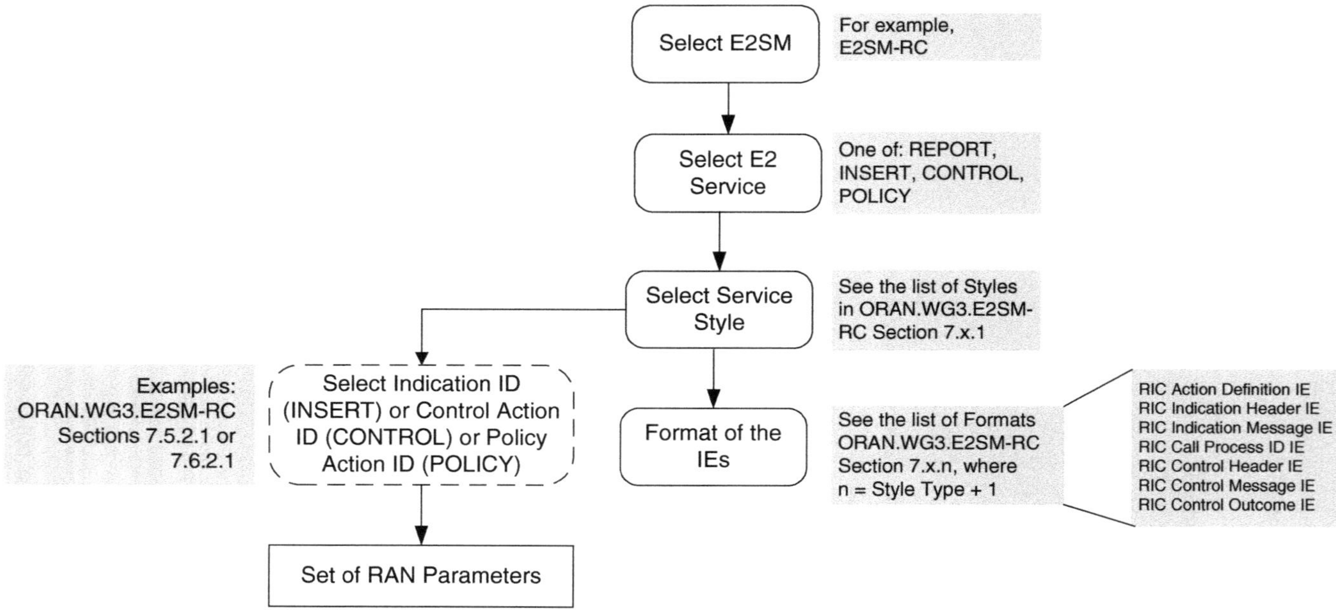

Figure 7.5 Selection of formats and RAN Parameters. (*Source:* Vladimir Yanover.)

Above Steps 1-4 determine the formats to be used for the Information Elements (IEs) and the set of RAN Parameters

The list of IE formats for Styles 1 to 4 can be found in Sections 7.4.2–7.4.5 correspondingly, in O-RAN.WG3.E2SM-RC-R003-v05.00 [9]. For example, 7.4.2 refers to formats, which shall be used in the case of REPORT Service Style 1.

7.5.2.2 Selection of IE Formats and RAN Parameters for Event Trigger Definitions

Selection of IE formats and RAN Parameters for Event Trigger definitions depends on several parameters:

- **Selection of the E2 Service Model (E2SM)**
  In all examples, the E2SM-RC is used
- **Selection of the RIC Event Trigger Style**; one of the following:
  - **Message Event**
    Transmission or reception of a message via RRC or one of the gNB network interfaces
  - **Call Process Breakpoint**
    Triggering Event is defined as entry to specific operation phase, described in 3GPP, such as Bearer Context Setup or Handover Preparation. For example, it can be that the E2 Node enters the Bearer Context Setup process, with the subscription that defines an Event Trigger with Style Type = 2 ("Call Process Breakpoint"), Call Process Type ID = 2, Call Break-point ID = 1, and Call Breakpoint Name = "Bearer Context Setup." Referred 3GPP specifications are 3GPP TS 38.410 [2], 3GPP TS 38.420 [16], 3GPP TS 38.460 [13], 3GPP TS 38.470 [14], 3GPP TS 38.473 [15].
  - **E2 Node Information Change**
    Triggering event occurs when E2 Node or cell configuration is changed
  - **UE Information Change**
    Triggering event occurs when UE Information changed

Figure 7.6 shows the process of selection; it should be noted that in the case of Event Trigger Style "Call Process Breakpoint," additionally selection of Call Process Type and Call Breakpoint ID should be performed.

Above selections determine the format to be used for the Event Trigger Definition IE and the set of RAN Parameters

7.5.3 Event Triggers

As described in Section 7.5.2.2 in the E2SM-RC, the Event Triggers are defined uniformly across E2 Services and Service Styles. For information of which Event

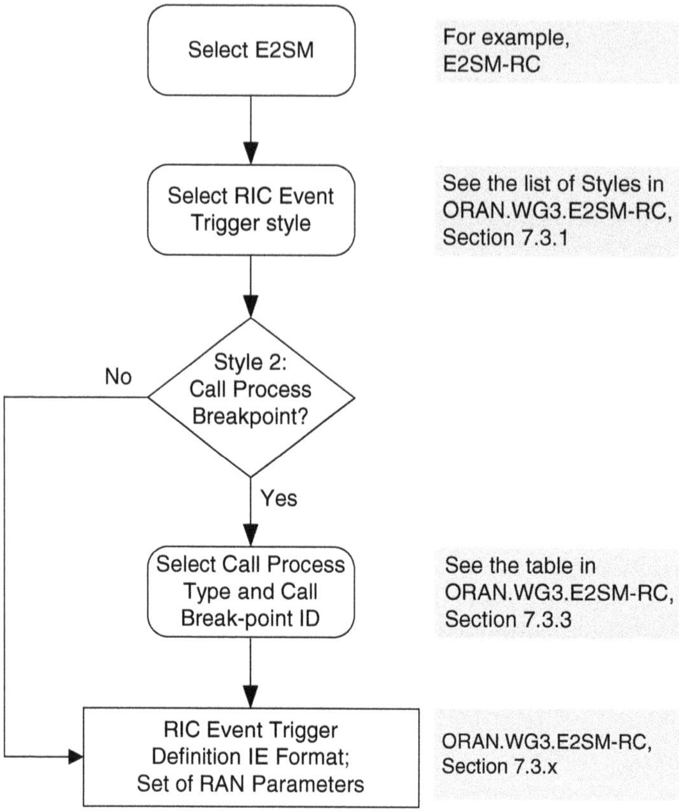

Figure 7.6 Event Trigger Definition: selection of formats and RAN Parameters. (*Source:* Vladimir Yanover.)

Triggers types are supported in specific E2 Services, see the corresponding Sections 7.5.4–7.5.8, and 7.3.1 in [9].

Four Styles of Event Triggers are supported.

- **Event Trigger Style 1**: "Message Event"
  This Style of Event Trigger is activated by incoming or outgoing message over one of E2 Node interfaces, such as Xn, NG, or RRC; see [2, 12, 16, 17]
- **Event Trigger Style 2**: "Call Process Breakpoint"
  This Style of Event Trigger is activated when the E2 Node enters one of predefined "call processes," such as Handover Preparation or UE Context Setup. The list of such processes is enumerated by Call Process Type ID (O-RAN.WG3.E2SM-RC-R003-v05.00 [9], Section 7.3.3):
  - Call Process Type ID = 1:UE Context Management
  - Call Process Type ID = 2: Bearer Context Management

- Call Process Type ID = 3:Mobility Management
- Call Process Type ID = 4: Multi-RAT Dual Connectivity Management
- Call Process Type ID = 5:Radio Resource Control Management
- Call Process Type ID = 6: PDU Session Management

Within every Call Process Type, multiple Call Breakpoints are defined, enumerated by the Call Break-point ID. For example, for the UE Context Management, the E2SM-RC specifies the following Call Breakpoints:
- Call Break-point ID = 1: UE Context Setup
- Call Break-point ID = 2: UE Context Modification
- Call Break-point ID = 3: UE Context Release

Similarly, under the Mobility Management, the following Call Breakpoints are defined:
- Call Break-point ID = 1: Handover Preparation
- Call Break-point ID = 2: Handover Cancel
- Call Break-point ID = 3: Handover Resource Allocation

- **Event Trigger Style 3**: "E2 Node Information Change"
  This Style of Event Trigger defines several call processes enumerated by E2 Node Information Change ID to differentiate between Cell Configuration Change, Cell Neighbor Relation Change, and Slice Configuration Change.
- **Event Trigger Style 4**: "UE Information Change"
  This Style of Event Trigger is activated by change in the information of specific UE. In this type of Event Trigger, the UE Identifier Change Type differentiates between New UE Connected, UE Handed Over, UE ID Changed, and UE ID Removed.

7.5.4 REPORT Service

7.5.4.1 Introduction

The REPORT service is used for collection of network data via the E2 interface.

The Near-RT RIC sets the REPORT service by using a RIC Subscription or RIC Subscription Modification procedure. The subscription sets the E2 Node to send the corresponding RIC INDICATION message after any occurrence of the Event Trigger defined in the subscription. Invocation of the REPORT service does not affect the associated procedure flow.

The content of the RIC INDICATION message is defined in the subscription as well. If the RIC INDICATION message is used in the REPORT service, the O-RAN specifications sometimes identify it as RIC INDICATION (REPORT) message.

7.5.4.2 Reported Network Data Categories

Usually, the RIC INDICATION (REPORT) message contains the values of network measurements, performed by the RAN, similarly to 3GPP Performance

Measurements (PM); see 3GPP TS 28.552 [18]. There is, however, some difference. 3GPP PM information is aggregated over the cell and over comparatively long time interval (minutes, hours). With the REPORT service, the Near-RT RIC can order and receive more granular information, in terms of both time and location. For example, it can be the information of the throughput of a single radio bearer sampled every 50 ms. Another category of information, available for collection via the REPORT service, is the results of measurement performed by the attached UEs and reported to the RAN via the RRC protocol. Yet another category is information of the state of internal RAN objects, such as DRBs and QoS Flows and of the state of individual UEs. Aggregation of such information over cells, RAN nodes, network slices is also supported.

There are two major E2 Service Models, which together cover needs in network data collection: E2SM-KPM and E2SM-RC.

7.5.4.3 REPORT Service Procedures

The REPORT service includes the following steps (Figure 7.7):

1) Near-RT RIC configures a RIC Subscription in the E2 Node; the subscription includes definition of the RIC INDICATION (Report) message to be sent after every occurrence of the Trigger Event. Usually the Trigger event is periodic.
2) During normal E2 Node functioning, the Event Trigger is detected

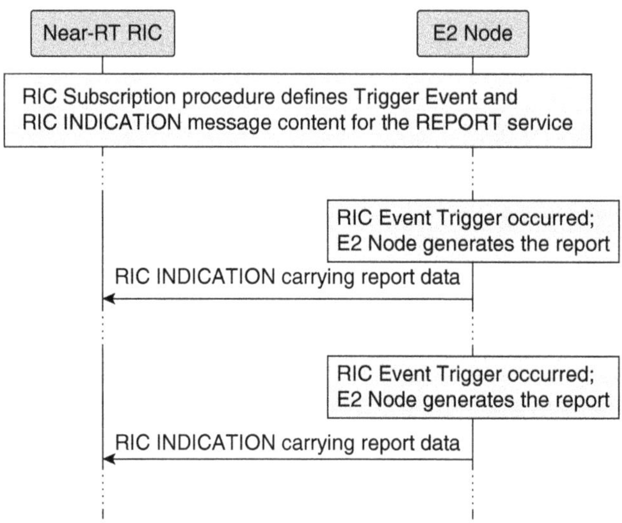

Figure 7.7 REPORT Service. (*Source:* Vladimir Yanover.)

3) The E2 Node sends the RIC INDICATION (Report) message to the Near-RT RIC; the message contains the REPORT information defined in the subscription
4) The E2 Node continues the associated procedure.

Steps 2–4 can be repeated while the subscription is active.

7.5.4.4 E2 Service Models for REPORT: E2SM-KPM and E2SM-RC

The E2SM-KPM (O-RAN.WG3.E2SM-KPM-R003-v04.00 [7]) enables periodic reporting of all kinds of performance measurements defined in 3GPP TS 28.552. On top of that, the O-RAN specifications allow collection of RAN measurements for a single UE, for multiple UEs (condition-based), for specific DRBs and QoS Flows, for slices, and more. Reports can be generated in short time intervals (milliseconds). More details can be found in Section 7.5.4.2.

The E2SM-KPM enables network data collection at the E2 Node; the E2 Node can be O-CU-CP, O-CU-UP, O-DU, or any combination of those. The E2 Node hosts the RAN Function "KPM Monitor," which receives a subscription from the Near-RT RIC (O-RAN.WG3.E2SM-R003-v05.00 [8], O-RAN.WG3.E2SM-KPM-R003-v04.00 [7]) and generates reports according to the subscription. REPORT is the only E2 Service supported in E2SM-KPM.

The E2SM-KPM supports one single Event Trigger Style: RIC Style Type 1, "Periodic Report," for which the RIC Event Trigger Definition IE contains the Reporting Period expressed in units of 1 millisecond.

E2SM-KPM supports the following REPORT Service Styles:

- **Style 1**: E2 Node Measurement
  With this Style, RIC INDICATION messages carry measurement report from the E2 Node
- **Style 2**: E2 Node Measurement for a single UE
  The RIC INDICATION message carries E2 Node Measurements for a single UE
- **Style 3**: Condition-based, UE-level E2 Node Measurement
  Same as Style 2, but for a group of UEs matching conditions defined in the subscription per measurement type
- **Style 4**: Common Condition-based, UE-level Measurement
  The RIC INDICATION message carries the measurement values for a group of UEs satisfying the condition
- **Style 5**: E2 Node Measurement for multiple UEs
  The RIC INDICATION message carries the measurement values for a group of UEs defined in the subscription

Example of format definition in E2SM-KPM: for the REPORT Service Style 1, E2SM-KPM defines the Action Definition IE Format 1; see Section 8.2.1.2.1 in O-RAN.WG3.E2SM-KPM-R003-v04.00 [7]. This format defines the parameters of data collection, which can be set in the subscription. One of parameters is Granularity Period, which specifies the collection interval for network measurements. The text string parameter Measurement Name identifies the measurements to be collected; it is done in line with 3GPP definitions of measurements in TS 28.552 [18]. Example of Measurement Name defined in 3GPP: DRB. UEThpDlDist.Bin.X (TS 28.552 [18], Section 5.1.1.3.2), where X denotes optional combination of subcounters representing QoS class, S-NSSAI, or PLMN ID. Use of the measurements names for identification of the collected measurements allows to reuse all measurement names specified in 3GPP. Mechanism of Measurement Names allows easy extension by adding new names.

In the E2SM-KPM, this extension mechanism is used for definition of O-RAN specific measurements with granularity of a single UE or even a single QoS Flow. The O-RAN Measurement Name is derived from the 3GPP Management Name by addition of the index (subcounter), which identifies specific UE or QoS Flow. For example, in the DRB.UEThpDlDist.Bin.X, the name component X can contain identification of specific UE or QoS Flow. Section 7.9.0 “Conversion of measurement definitions” in [7] explains this mechanism. For example, for the Throughput category of measurements defined in TS 28.552 [18], the O-RAN specification prescribes use of same throughput formulas as the 3GPP specification but restricted to specific UE or QoS Flow.

The same extension mechanism is used for definition of O-RAN specific Measurement Names, which are not derived from 3GPP Measurement Names; see Section 7.10 “O-RAN specific Performance Measurement” in the specification O-RAN.WG3.E2SM-KPM-R003-v04.00 [7]. Similarly to 3GPP specifications, O-RAN specific Measurement Names may include subcounters, which specify the scope where the measurements are generated, for example, to Slice ID or specific 5QI or specific QFI.

For the RIC INDICATION Message, where the actual measurement values are transferred, the E2SM-KPM defines Indication Message Format 1 (O-RAN.WG3. E2SM-KPM-R003-v04.00 [7], Section 8.2.1.4.1). The RIC INDICATION message links to the corresponding subscription by use of same Measurement Names and labels (subcounters) as in the subscription.

In addition to the E2SM-KPM, the E2SM-RC model O-RAN.WG3.E2SM-RC-R003-v05.00 [9] supports the following REPORT Styles:

- **Message copy**
  RIC INDICATION message carries the content of a 3GPP-defined message, received or transmitted by the E2 Node
- **Call Process Outcome**
  RIC INDICATION message carries information of the outcome of the process defined in the “Call Process Breakpoint” Event Trigger

- **E2 Node Information**
  RIC INDICATION message carries the E2 Node information or cell level information
- **UE Information**
  RIC INDICATION message carries UE related state and performance information in the context defined in the "UE Information Change" Event Trigger

For the REPORT Service, the E2SM-RC defines the following Event Trigger Styles:

- For the Service Style = 1: Event Trigger Style = 1 "Message Event"
- For the Service Style = 2: Event Trigger Style = 2 "Call Process Breakpoint"
- For the Service Style = 3: Event Trigger Style = 3 "E2 Node Information Change"
- For the Service Style = 4: Event Trigger Style = 4 "UE Information Change"

See 7.5.3 for more details on Event Trigger Styles.

7.5.5 INSERT Service

7.5.5.1 Introduction

With the INSERT service, Near-RT RIC can receive near-real-time notifications of the RAN events that occurred in the E2 Node. In this section, the focus is on the scenario when the INSERT service action is coupled with immediately following CONTROL service action. The role of the CONTROL service is to provide near-real-time reaction to the notification carried by the INSERT Indication message. In such scenarios, the Near-RT RIC becomes involved in the corresponding near-real-time control loops, which is an important building block of RAN programmability.

Near-RT RIC sets the INSERT service by using RIC Subscription (or RIC Subscription Modification) procedure. The subscription requests that after each Event Trigger, the E2 Node suspends the associated procedure and sends a RIC INDICATION (Insert) message to the Near-RT RIC. The Event Trigger and the format of the RIC INDICATION (Insert) message are defined in the subscription. Suspension of the associated procedure is an important aspect of the INSERT service.

Functional aspects of the INSERT/CONTROL process depend on the E2 Service Model. Specifically, in the E2SM-RC model, the E2 Node decides WHAT to do, while the role of the Near-RT RIC is to decide HOW to do it. For example, after reception of an RRC MeasurementReport message from a UE, which informs the Near-RT RIC on the A3 RAN event (3GPP TS 38.331 [10]), the E2 Node (O-CU) may decide that handover (HO) is in order. If appropriate subscription is in place, the E2 Node sends the INSERT Indication message requesting the Near-RT RIC to provide HO details, such as selection of the HO target cell(s) and/or (re)configuration of HO parameters.

7.5.5.2 INSERT Service Procedures

In the scenario where the INSERT service action is coupled with immediately following CONTROL service action, the process includes the following steps (Figure 7.8):

1) The Near-RT RIC sets the subscription, using the RIC Subscription procedure. The subscription defines the Event Trigger and contains information on the required actions, such as RIC Action Definition, and on the subsequent actions, such as the type of the RIC Subsequent Action and RIC Time to Wait.
2) When an Event Trigger is detected, the E2 Node suspends the associated procedure and starts Time to Wait timer.
3) The E2 Node sends RIC INDICATION (Insert) message to the Near-RT RIC; the message may contain identification of the suspended instance of the associated procedure and information on the actions selected by the E2 Node from the list of RIC Action Definitions in the subscription. Such information essentially is a request to the Near-RT RIC for instructions.
4) After reception of the RIC INDICATION (Insert) message, the Near-RT RIC may decide to intervene in the process and issue a RIC CONTROL REQUEST message, which instructs the E2 Node on further actions.

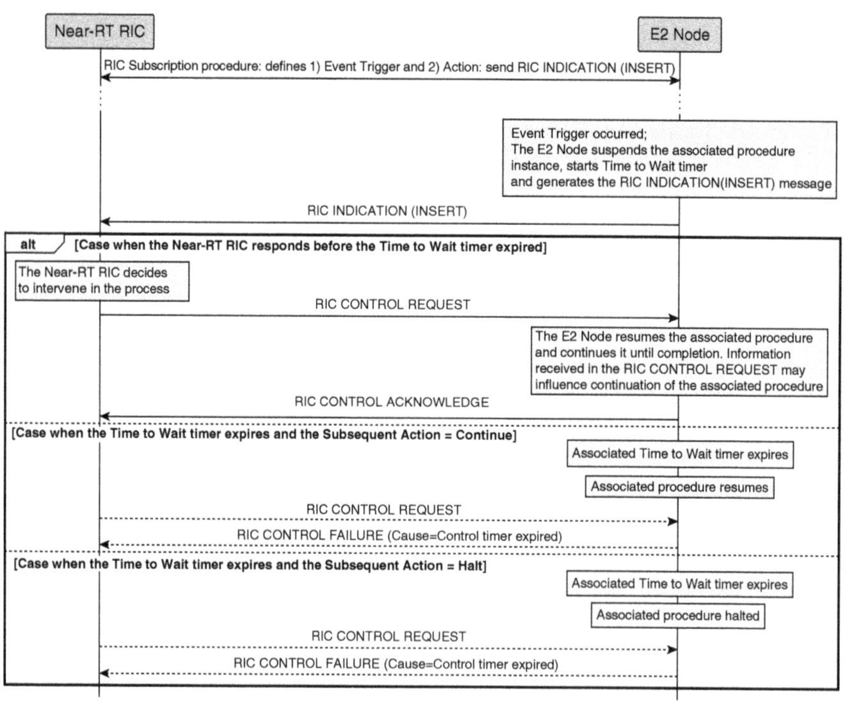

Figure 7.8 RIC INSERT Service coupled with RIC CONTROL action. (*Source:* Vladimir Yanover.)

5) The E2 Node receives the RIC CONTROL REQUEST message. Reception of the message resumes the suspended associated procedure. The E2 Node identifies the suspended associated procedure according to information in the RIC CONTROL REQUEST, such as Call Process ID. The instructions carried in the RIC CONTROL REQUEST message may influence execution of the associated procedure, steering it to the desired direction. For example, in the RAN Control E2 Service Model (E2SM-RC) [9], the Service Style 1 "Radio Control," allows the Near-RT RIC to use the RIC CONTROL REQUEST message with Control Action ID = 1 to indicate the Handover Control and specify the handover target with the parameter Target Primary Cell ID.

At the step 5, further E2 Node actions depend on whether the E2 Node receives the RIC CONTROL REQUEST message before or after expiry of the Time to Wait timer:

- if the E2 Node receives the RIC CONTROL REQUEST message before expiry of the Time to Wait timer, the E2 Node resumes the associated procedure, using the information from the RIC CONTROL REQUEST message.
- if the Time to Wait timer expires before the RIC CONTROL REQUEST message received and the Subsequent Action Type was set to Continue in the subscription, the E2 Node continues the original associated procedure.
- if the Time to Wait timer expires before the RIC CONTROL REQUEST message received and the Subsequent Action Type was set to Halt, the E2 Node halts the associated procedure.

If the E2 Node receives the RIC CONTROL REQUEST message after Time to Wait timer expiry, the E2 Node sends RIC CONTROL FAILURE message with failure Cause identified as "Control timer expired."

Steps 2–5 can be repeated while the subscription is active.

7.5.5.3 E2SM-RC Service Model for INSERT

The E2 INSERT Service supports for Service Styles 1–7 only one Event Trigger Style:

- **Event Trigger Style 2**: Call Process Breakpoint
  This Event Trigger is activated when the E2 Node enters one of predefined "Call Processes," enumerated by Call Process Type ID and proceeds to one of "Call Process Breakpoints" enumerated by Call Break-point ID; see Section 7.5.3.
  The following service styles for INSERT are supported by the E2SM-RC model:

- **Style 1**: Radio Bearer Control Request is used for requesting modification of the parameters, related to the Radio Bearer Control such as 5QI, Packet Delay Budget, and Maximum/Guaranteed Flow Bit Rate Downlink/Uplink
- **Style 2**: Radio Resource Allocation Control Request is used for requesting modification of the Radio Resource Allocation parameters; examples:

resourceAllocation, sps-Config configuration parameters, and cqi-Table defined in 3GPP TS 38.331 [10].

- **Style 3**: Connected Mode Mobility Control Request is used for triggering of a CONNECTED state mobility procedure, such as Handover, Conditional Handover (CHO), and Dual Active Protocol Stack (DAPS) for specific UE, toward specified handover target cell(s)
- **Style 4**: Radio Access Control Request is used for modification of the parameters controlling UE access to the gNB cells
- **Style 5**: Dual Connectivity Control Request is used for requesting initiation of dual connectivity (DC) actions, such as DC Secondary Node addition
- **Style 6**: Carrier Aggregation Control Request is used for requesting initiation of carrier aggregation (CA) actions, such as CA Secondary Cell addition or release
- **Style 7**: Idle Mode Mobility Control Request is used for modification of IDLE state mobility related parameters, such as cell re-selection priority
- **Style 255**:Multiple Actions Control Request is used for requesting multiple parallel actions

In all styles, the E2 Node decides WHAT to do while the role of Near-RT RIC is to decide HOW to do it. Usually INSERT Indication message queries the Near-RT RIC about modification of the corresponding parameters. For example, for the Style 3 "Connected Mode Mobility Control Request," the E2SM-RC prescribes use of the INSERT Indication message Format 5, which contains the "List of RAN parameters requested." It means that the INSERT Indication message is requesting the Near-RT RIC to respond and provide the values for these parameters. The corresponding parameters are identified in Section 8.4.4 of O-RAN.WG3.E2SM-RC-R003-v05.00 [9]. For these parameters, the E2SM-RC specification refers to the definitions done in 3GPP. Two examples of such references:

- **Target Cell**
  The TS 38.423 [17], clause 9.2.3.25, defines Target Cell Global ID for use in the Xn-based handover. For an NR target cell, it's NR CGI.
- On the DRBs for handover, e.g. DRBs to Be Setup
  The TS 38.473 [15], clause 9.2.2.1, defines these parameters for use in the F1AP protocol.

For every INSERT Style, specification [9] describes Action Definition IE format. For example, for Style 3 "Connected Mode Mobility Control Request," the E2SM-RC defines Action Definition IE format in Section 9.2.1.2.3 "E2SM-RC Action Definition Format 3" of [9]. This IE contains the List of RAN parameters and optionally UE ID. If the UE ID included, the INSERT actions shall be applied to specific UE only.

7.5.6 CONTROL Service

7.5.6.1 Introduction

The CONTROL Service enables Near-RT RIC to control operations of the E2 Node. There are two important use scenarios for the CONTROL service:

- CONTROL coupled with prior INSERT Service invocation
- "Standalone" CONTROL, not coupled with INSERT (also termed as "RIC Service CONTROL initiated by NEAR-RT RIC").

The first scenario is described in Section 7.5.5 with the process flow shown in Figure 7.8.

In the second, "standalone," scenario (Figure 7.9), the Near-RT RIC determines the desired way for further E2 Node operations. If this requires execution of specific actions and/or change in the E2 Node configuration, the Near-RT RIC sends to the E2 Node the corresponding instructions in a RIC CONTROL REQUEST message. Here are two examples of such instructions:

- Configure mobility parameters for specific UEs, such as the thresholds of An/Bn events. This part belongs to CONTROL Service Style 9 "Measurement Reporting Configuration Control" (O-RAN.WG3.E2SM-RC-R003-v05.00 [9], Section 8.4.10)

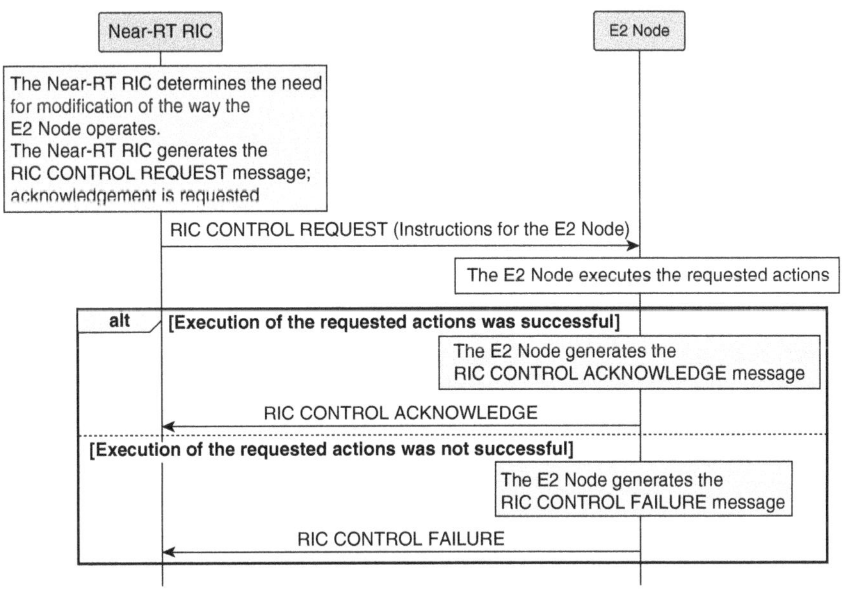

Figure 7.9 "Standalone" RIC CONTROL REQUEST. (*Source:* Vladimir Yanover.)

- Modify RAN configuration parameters, such as "List of Secondary cells to be setup," in the UE admission control part of CONTROL service Style 4 "Radio Access Control" (O-RAN.WG3.E2SM-RC-R003-v05.00 [9], Section 8.4.5.1).

In the "standalone" scenario, the CONTROL REQUEST message does not indicate previously suspended associated procedure; it may include Control Ack Request.

The E2 Node receives the RIC CONTROL REQUEST message and executes the requested actions. The E2 Node sends acknowledgement if requested. If the execution was successful, it's RIC CONTROL ACKNOWLEDGE message, otherwise RIC CONTROL FAILURE message.

7.5.6.2 E2 Service Models for CONTROL

The E2SM-RC model supports the following Service Styles for E2 CONTROL Service. All these Styles can be used in both scenarios: a CONTROL coupled with INSERT and a "standalone" CONTROL.

- **Style 1**: Radio Bearer control; it is used for modification of the parameters related to radio bearers and/or modification of the E2 Node behavior for specific UE. This includes modification of the following: DRB QoS parameters, mapping between QoS Flows and DRBs, Logical Channel configuration, configuration of the Radio bearer admission control, configuration of the Split bearer
- **Style 2**: Radio resource allocation control; this Style is used for modification of Radio Resource Management related parameters and/or of behavior of the E2 Node for specific scope. The scope may be the whole E2 Node, cell, slice, specific UE, QoS Flows or DRBs with specific QoS category. The related parameters belong to such functions as Discontinuous Reception (DRX), Scheduling Request (SR), Configured Grant, Semi-Persistent Scheduling (SPS), Slice level PRB quota, and Channel Quality Indicator (CQI) table
- **Style 3**: Connected mode mobility control Style; it is used for configuration of a CONNECTED state mobility procedure (Handover or Conditional Handover). Dual Active Protocol Stack (DAPS) is supported as well
- **Style 4**: Radio access control Style; it is used for modification of parameters and/or behavior of the E2 Node in the part of UE access to cells. The related functions are RACH back-off, UE admission, RRC connection reject, Access barring, and RRC connection release
- **Style 5**: Dual connectivity (DC) control Style; it is used for control of Dual Connectivity operations including change of bearer type or change of the bearer termination point between the MN and SN
- **Style 6**: Carrier Aggregation (CA) control Style; it is used for control of Carrier Aggregation operations
- **Style 7**: Idle mode mobility control Style; it is used for control of IDLE state operations, such as modification of the cell reselection priority and idle timers

for cell reselection in same frequency channel or between frequency channels, and between different RATs

- **Style 8**: UE information and assignment Style; it is used for control of association of the UE with a RAN UE group, in particular, for POLICY services support
- **Style 9**: Measurement Reporting Configuration control Style; it is used for control of the measurement report configuration
- **Style 10**: Beamforming Configuration control; it is used to set or modify beamforming configuration toward specific UE.
- **Style 255**: Multiple Actions Control Style

As described in Section 7.5.2.1, within given Service Style, the Control Action ID is used to indicate specific actions requested by the Near-RT RIC and selection of the RAN Parameters. For example, for the Service Style 1 "Radio Bearer control," the following Control Actions are defined in the E2SM-RC:

- **DRB QoS Configuration**: it is used for control of the DRB QoS configuration parameters
- **QoS flow mapping configuration**: it is used for control of the mapping of QoS flows to a DRB
- **Logical channel configuration**: it is used for control of the Logical Channel ID (LCID) configuration of a DRB
- **Radio admission control**: it is used for control of RAN admission toward specific UE
- **DRB termination control**: it is used for control (change) of the bearer termination point
- **DRB split ratio control** : it is used for control of the DRB split between its RLC entities
- **PDCP Duplication control** : it is used for configuration of the number of legs, the number of RLC entities in the DRB, and of the PDCP duplication

Table 7.4 provides some examples of RAN parameters for CONTROL Service Style = 1, Control Action ID = 1 ("DRB QoS Configuration"). Section 8.4.2.1 in O-RAN.WG3.E2SM-RC-R003-v05.00 [9] specifies the RAN parameters, which can

Table 7.4 Examples of RAN parameters for CONTROL Service Style = 1, Control Action ID = 1

RAN parameter	RAN parameter definition in 3GPP
5QI	TS 37.483 [11], clause 9.3.1.27 or 9.3.1.28
Packet Delay Budget	TS 37.483 [11], clause 9.3.1.47
Packet Error Rate	TS 37.483 [11], clause 9.3.1.48

be modified by the E2 Node per request from the Near-RT RIC. The table refers to E1AP specification TS 37.483 [11].

7.5.7 POLICY Service

7.5.7.1 Introduction

The POLICY Service is used for setting the policy via the E2 interface. The policy defines a set of actions to be executed by the E2 Node after certain trigger, with optional notification to the Near-RT RIC. The POLICY functionality allows programming of internal real-time control loops in the E2 Node.

The POLICY Service definitions assume that for every trigger there is associated procedure for handling this trigger, sometimes this procedure is called "default RAN behavior." The effect of the policy (Policy Action) can be modification or complete replacement of the "default RAN behavior." The definition of every specific policy includes specifications of the Event Triggers, with one or more of Policy Actions associated with every Event Trigger. For example, in the POLICY Service Style 3 "Connected Mode Mobility Control," Policy Action ID = 1 corresponds to the Policy for Handover Control, which includes setting up a policy for control of the handovers of the primary cells. One of possible parameters used for such control is Target Primary Cell ID.

The E2SM-RC model supports several Service Styles for POLICY, with further functional subdivision, based on Policy Action ID (Section "E2SM-RC Policy Actions"). For every Style/Policy Action ID, there are two Policy Approaches:

- **Control Policy Approach**: which is defined as a couple {Policy Condition, Policy Action}, using same data structure as CONTROL Service (see Section 7.5.7.4) to replace the default RAN behavior (associated procedure)
- **Offset Policy Approach**: which is defined with a set of multiple couples {Policy Condition, Policy Action}; see Section 7.5.7.5. It is used to modify default RAN behavior (associated procedure)

7.5.7.2 POLICY Service Procedures

The POLICY service procedure flow includes the following steps (Figure 7.10):

1) (Optional) Near-RT RIC applies a dedicated Subscription procedure with REPORT action to provide information on the outcome of the associated procedure after the POLICY actions are executed.
2) Near-RT RIC applies the RIC Subscription procedure to configure a policy at the E2 Node. The subscription includes definition of:
   - The Event Trigger
   - The Policy Conditions
   - The Policy Action to be performed by the E2 Node after each Event Trigger

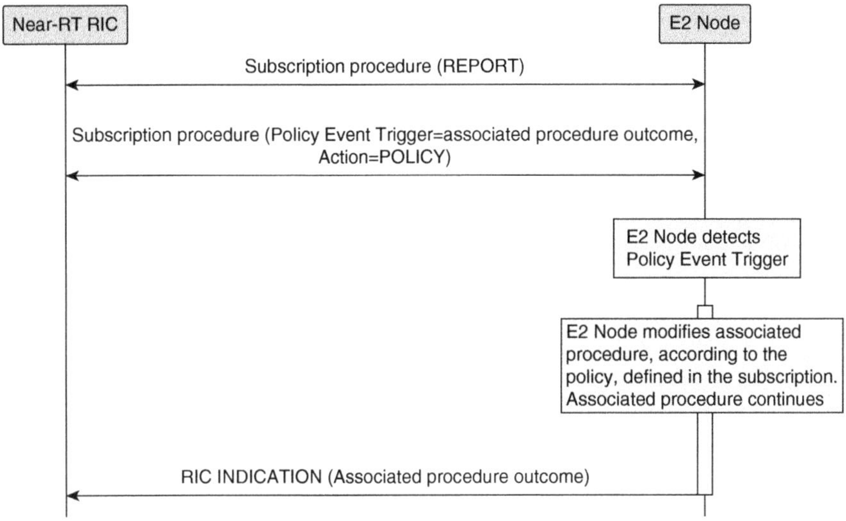

Figure 7.10 RIC POLICY Service. (*Source:* Vladimir Yanover.)

3) During operations of the E2 Node, the Event Trigger is detected.
4) The E2 Node determines further actions, according to the Policy Action information. If the Policy Conditions are not defined in the subscription or they are defined and met, the action is executed, which can result in modification of the ongoing associated procedure or cancellation of the associated procedure and initiation of different procedure. If the subscription defined the Policy Conditions, but they are not met, no actions; in this case, the default RAN behavior is in order.
5) In the E2 Node, execution of the associated procedure instance continues until completion
6) If the Step 1 has been executed, the E2 Node sends a REPORT Indication message with information on the outcome of the associated procedure; see about the "Call Process Outcome" in Section 7.5.4.2.

7.5.7.3 E2SM-RC (RAN Control) POLICY Model

7.5.7.3.1 Policy Styles The following Service Styles for POLICY are supported in the E2SM-RC model.

- **Style 1**: Policy for Radio Bearer Control (RBC); it is used to define a policy, which can modify the behavior of the Radio Bearer Control procedure for specific RAN UE Group. The term "RAN UE Group" means grouping of UEs based on UE list or on one or more of the UE properties related to cell, slice, QoS.

- **Style 2**: Policy for Radio resource allocation control; it is used to define a policy, which can modify the behavior of Radio Resource Allocation control for specific RAN UE Group
- **Style 3**: Policy for Connected mode mobility control; it is used to define a policy, which can modify the behavior of connected mode mobility procedures (HO or CHO), for a specific RAN UE Group. The Style can be applied also to the DAPS
- **Style 4**: Policy for Radio Access control; it is used to define a policy, which can modify the behavior of Radio Access control procedures, such as RACH backoff, UE Admission, RRC connection admission and release, Access barring,
- **Style 5**: Policy for Dual connectivity (DC) control; it is used to define a policy, which can modify the behavior of DC procedures, when applied to specific RAN UE Group
- **Style 6**: Policy for Carrier Aggregation (CA) control; it is used to define a policy, which can modify the behavior of carrier aggregation (CA) procedures, when applied to specific RAN UE Group
- **Style 7**: Policy for Idle mode mobility control; it is used to define a policy, which can modify the behavior of idle mode configuration procedures, when applied to specific RAN UE Group
- **Style 8**: Policy for Measurement Reporting Configuration Control; it is used to define a policy for measurements configuration for the UE measurements reports defined in the TS 38.331 [10]
- **Style 9**: Policy for Beamforming Configuration control; it is used for setting a policy for beamforming configuration used for individual UE

7.5.7.3.2 Event Triggers As described in Section 7.5.2.2 in the E2SM-RC, the Event Triggers are defined uniformly across E2 Services and Service Styles. For the POLICY Service, three Styles of Event Triggers are supported:

- **For POLICY Service Styles 1–7**: Event Trigger Style 1 "Message Event"
- **For POLICY Service Styles 1–8**: Event Trigger Style 2 "Call Process Breakpoint"
- **For POLICY Service Style 9**: Event Trigger Style 4 "UE Information Change"

7.5.7.3.3 E2SM-RC Policy Actions Policy Action description consists of one or several couples {Policy Condition definition, Policy Action definition}, encoded with:

- **(Optional) Policy Condition Definition IE**
  The test to be applied as a condition for Policy Action execution
- **Policy Action Definition IE**
  Defines policy action to be applied; execution of the action is conditioned by the outcome of the test specified in the Policy Condition Definition IE (if present)

Programming of the Policy Actions enables modification (steering) of the associated procedure, dedicated to handling the corresponding Event Trigger. Such modification may include change in the internal configuration of the corresponding RAN functionalities, such as DRB QoS control, Radio bearer admission control, operations related to handover and Dual Connectivity (DC). The list of the corresponding Styles can be found in Section 7.5.7.3.1 "Policy Styles".

After occurrence of the Event Trigger, the E2 Node checks the list of Policy Actions defined in the subscription. The Policy Action description uses the RIC Action Definition IE Format 2, and therefore may optionally contain Policy Condition Definition IE (O-RAN.WG3.E2SM-RC-R003-v05.00 [9], Section 9.2.1.2.2).

The E2 Node goes through the list of Policy Actions, evaluating the associated Policy Conditions to determine which Policy Action(s) should be executed, if any.

Execution of the Policy Actions depends on the chosen mode of operations called "Policy Approach." The E2SM-RC defines two modes of POLICY operations (see Section 7.5.7.1):

- Policy Approach "Control," in which the Policy Action replaces the default RAN behavior
- Policy Approach "Offset," in which the Policy Action modifies the default RAN behavior, usually by modification of the parameters, which control the behavior, e.g. adding an offset to a threshold.

The syntax and semantics of the Policy Conditions can be outlined as follows.

For every POLICY Style, O-RAN.WG3.E2SM-RC-R003-v05.00 [9] specifies the RAN Parameters, which can be used in POLICY Conditions. Section 8.5.1A in the referred document [9] specifies the common set of RAN Parameters used for all Service Styles. Sections 8.5.2–8.5.10 in [9] identify the RAN Parameters used for Styles 1–9 correspondingly. In addition, Section 8.1 in the referred document [9] defines the RAN parameters for Event Triggers, across all Styles.

To specify a POLICY Condition, the selected RAN Parameters are combined into a RAN Parameter Test IE (O-RAN.WG3.E2SM-RC-R003-v05.00 [9], Section 9.3.30), which describes recursively defined hierarchical logical expression composed of

- **RAN Parameter Test Condition**: allows to include into logical expression such elements as Comparison, with such possible values as "equal," "difference," "greaterthan," and Presence with such possible values as "present," "configured"
- **RAN Parameter Value**: one of: BOOLEAN, INTEGER, REAL, BIT STRING, OCTET STRING, PrintableString
- **Logical OR**: if "TRUE," the operation applied to the next condition is "OR." Otherwise, it's "AND."

Under each of POLICY Service styles, multiple possible Policy Actions are defined. For example, under the POLICY Service Style 3: Connected Mode Mobility Control, the following Policy Actions are specified (O-RAN.WG3.E2SM-RC-R003-v05.00 [9], Section 7.7.4):

- **Policy Action ID = 1, Policy for Handover Control**: used for setting a policy for the selection of the primary cell and secondary cell during handover
- **Policy Action ID = 2, Policy for Conditional Handover Control**: used for setting a policy for the CHO in the part related to target primary cells
- **Policy Action ID = 3, Policy for DAPS Handover Control**: used for setting a policy for the DAPS handover applied to source and target cells.

For use with Connected Mode Mobility Control, the E2SM-RC model defines a common set of RAN Parameters (O-RAN.WG3.E2SM-RC-R003-v05.00 [9], Section 8.5.4.0), across all Policy Action IDs. Examples are:

- NR CGI, global NR cell identifier
- Reported NR RRC Measurements, defined in TS 38.331 [10] as a part of the measResult IE.

In addition, for every Policy Action, the E2SM-RC identifies specific set of RAN parameters. For example, in the case of Policy Action "Policy for Handover Control," it is a set of "offset" parameters in O-RAN.WG3.E2SM-RC-R003-v05.00 [9], Section 8.5.4.1.2. Examples of "offset" parameters for Policy for Handover Control:

A3 Event Threshold offset structure

- RSRP Offset
- RSRQ Offset
- SINR Offset
- Hysteresis

These parameters correspond to the following IEs in the TS 38.331 [10]:

- RSRP-RangeIE
- RSRQ-Range IE
- SINR-Range IE
- Hysteresis IE

7.5.7.4 Policy Approach "Control" Example

The following example describes a policy for handover execution; Service Style = 3 "Policy for Connected mode mobility control." In this example, the associated procedure responds to the RRC MeasurementReport message, received from specific UE, by triggering A3 RAN event (TS 38.331 [10]). The associated procedure's action is unconditional initiation of the UE handover to the corresponding neighbor cell. The Event Trigger is defined as arrival of RRC MeasurementReport message: Event Trigger Style 1 "Message event."

The Policy Condition is defined as follows: *(RSRP of the neighbor cell reported in the RRC message> “x”) AND (The neighbor cell load< “y”) AND (HO success rate to the neighbor cell > “z”),* where the threshold parameters x, y, z are set in the subscription.

The Policy Action is defined according to Policy Action ID = 1 “Policy for Handover Control,” which may contain RAN parameters such as the List of QoS flows in the DRB.

7.5.7.5 Policy Approach “Offset” Example

This example is about the Connected Mode Mobility call flow for which the RIC POLICY “Offset” approach is used. The policy provides guidance for handling of RRC Measurement reports received from a UE connected to the E2 Node (O-gNB or O-CU). The policy mechanism provides differentiation in handover parameters for selected individual UEs.

Figure 7.11 shows the process flow. It starts from RIC Control procedure, used for assignment of specific UEs to the “Explicit UE list.” Next step is the Subscription procedure, setting the RIC Policy with the following properties:

- The RIC Event Trigger Definition with Event Trigger Style = Call Process Breakpoint, Call Process Type = Mobility Management, Call Breakpoint

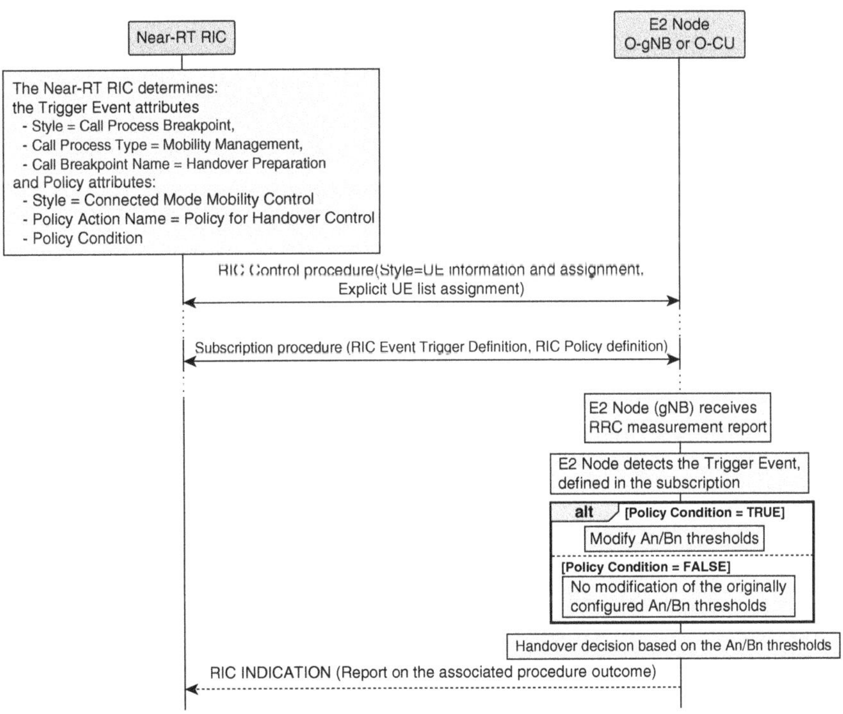

Figure 7.11 RIC POLICY example. (*Source:* Vladimir Yanover.)

Name = Handover Preparation. It is assumed that when the E2 Node (O-gNB or O-CU) receives an RRC Measurement report, the associated procedure goes to the Handover Preparation breakpoint. The Event Trigger Definition RAN parameters identify the UEs, to which the Policy is applied. In this example, it's RAN parameters identifying the value of the Index to RAT/Frequency Selection Priority (IRFSP; see O-RAN.WG3.E2SM-R003-v05.00 [8], Section 6.2.3.27 and TS 38.413 [12], clause 9.3.1.61).

- RIC Policy definition, using Style = Connected Mode Mobility Control, Policy Action Name = Policy for Handover Control. The Policy Condition Definition includes RAN parameters referring to the Serving Cell and to Explicit UE List membership. The Policy Action Definition parameters define a set of numerical offsets to be applied to the thresholds configured in the E2 Node for An/Bn RAN events (3GPP TS 38.331 [10]) before the thresholds are used for handover decision.

After the E2 Node determined the handover thresholds, with or without offsets, handover procedure continues: the handover thresholds are applied to specific signal measurement values, reported in the RRC message, which results in the handover decision.

7.5.8 QUERY Service

O-RAN specification O-RAN.WG3.E2GAP-R003-v05.00 [5] defines the QUERY service, which aims transfer of information from an E2 Node to the Near-RT RIC. This information can be related to RAN nodes or to the UEs registered at the RAN nodes.

The QUERY Service is based on RIC QUERY REQUEST, RIC QUERY RESPONSE, and RIC QUERY FAILURE messages. The process flow starts when the Near-RT RIC determines a need for pulling certain information from an E2 Node. Then the Near-RT RIC sends RIC QUERY REQUEST, which contains identification of the target RAN Function and definition of the query in the RIC Query Definition IE. The E2 Node delivers the requested information in the message RIC QUERY RESPONSE or returns RIC QUERY FAILURE in case of failure.

7.6 Other E2 Service Models: E2SM-CCC

The E2 Service Model Cell Configuration and Control (E2SM-CCC), specified in O-RAN.WG3.E2SM-CCC-R003-v03.00 [6], aims reading and setting the node level and cell level configuration in the E2 Node. The E2 Node is assumed to host the RAN Function "Cell Configuration and Control," which supports two E2

Services. The REPORT Service is used for reading node/cell level configuration parameters; the CONTROL Service used for setting/modifying these parameters.

For the REPORT Service, the E2SM-CCC specification defines two Service Styles: Styles 1 and 2 to report "Node-Level Configuration" and "Cell-Level Configuration" correspondingly.

For the REPORT Service Style 1, the E2SM-CCC specifies RIC Action Definition IE Format 1, which allows selection between unconditional reporting and reporting in the case of change. This format identifies the attributes to be reported, by Attribute Name. Dependently on the type of E2 Node (O-CU, O-DU, . . . etc.), the set of reported attributes may include identification of the involved entities, RRM Policy parameters, and some others. The corresponding RIC Indication Message IE Format 1 contains the reported values of the attributes.

For the REPORT Service Style 2, the E2SM-CCC specifies RIC Action Definition IE Format 2, which is similar, but oriented to cell level configurations.

For the CONTROL Service, the E2SM-CCC specification defines Style 1 "Node Configuration and Control" and Style 2 "Cell Configuration and Control."

For the CONTROL Service Style 1, the E2SM-CCC specifies RIC Control Message IE Format 1, which provides the old and new attribute values for the E2 Node Configuration.

Similar structure is used for the CONTROL Service Style 2 for the cell configuration attributes. The set of attributes may include configuration parameters specified in 3GPP TS 28.541 [19].

7.7 Example of CONNECTED State Mobility Solution with INSERT/CONTROL

Annex Section A.2 in O-RAN.WG3.E2SM-RC-R003-v05.00 [9] describes a solution, based on a combination of E2 INSERT Service and E2 CONTROL Service; the E2 INSERT Service is using the Connected Mode Mobility Event Trigger. The process includes that the E2 Node calls the Near-RT RIC for guidance when the E2 Node receives an RRC MeasurementReport message with a neighbor cell measurement matching one of configured HO conditions, such as A3 RAN event ("Neighbor becomes offset better than SpCell," 3GPP TS 38.331 [10]). It is assumed that the associated procedure in such case is to prepare a HO to the neighbor cell. Intervention of the Near-RT RIC in the handover process, e.g. in selection of the HO target, enables intelligent handover decisions, e.g. for traffic steering from overloaded cells to less loaded cells.

In the INSERT/CONTROL-based implementation, the Event Trigger for E2 INSERT is defined with Style = 2 (Call Process Breakpoint). Specific Call Process Breakpoint is identified by the Type = 3 (Mobility Management) and the

Call Breakpoint ID = 1 (Handover Preparation); see Section 7.3.3 in O-RAN.WG3.E2SM-RC-R003-v05.00 [9]. The subscription for E2 INSERT Service is defined with INSERT Service Style = 3 (Connected Mode Mobility Request), Indication ID = 1 (Handover control request). Such configuration implies use of RIC Action Definition Format 3, which allows the E2 Node to specify HO control parameters (O-RAN.WG3.E2SM-RC-R003-v05.00 [9], Section 8.4.4.1 Handover Control), such as: Target Cell, List of DRBs and PDU sessions for handover, every session with a List of QoS flows, a list of secondary cells, etc.

Figure 7.8 shows the process flow. The process starts when a UE in CONNECTED state sends an RRC MeasurementReport message with a neighbor cell measurement matching one of configured HO conditions, such as A3 RAN event. The E2 Node starts the associated procedure and makes a preliminary handover decision. Then the E2 Node suspends the associated procedure and, as a part of E2 INSERT service procedures, sends to the Near-RT RIC an RIC Indication message, which contains attributes of this decision, such as selected target primary cell and the list of retained DRBs.

The Near-RT RIC performs analysis of the information received in the RIC INDICATION message, makes its own decision about the planned HO, and sends the RIC CONTROL REQUEST message to the E2 Node. The format of the RIC CONTROL REQUEST message corresponds the Control Style = 3 (Connected Mobile Mobility) and Control Action ID = 1 (Handover control). Such format allows to confirm or override the parameters of preliminary HO decision received in the INSERT Indication message: Target Cell, List of DRBs and PDU sessions for handover, etc. For example, the Near-RT RIC may decide to down select retained DRBs or QoS Flows.

In general case, when the E2 Node receives the RIC CONTROL REQUEST before the Time to Wait timer expired, the E2 Node releases and executes the original or modified associated procedure.

7.8 Near-RT RIC Internal Architecture, xApps

7.8.1 Introduction

Figure 7.12 outlines the Near-RT RIC internal architecture, including facilities that support hosting of xApps. The xApps are third-party applications running on the Near-RT RIC platform (O-RAN.WG3.RICARCH-R003-v05.00 [20]). The figure shows the following architecture components:

- Near-RT RIC Database, which contains RAN and UE information. Using Shared Data Layer (SDL), xApps can read and write this information; they can also subscribe to database notification services

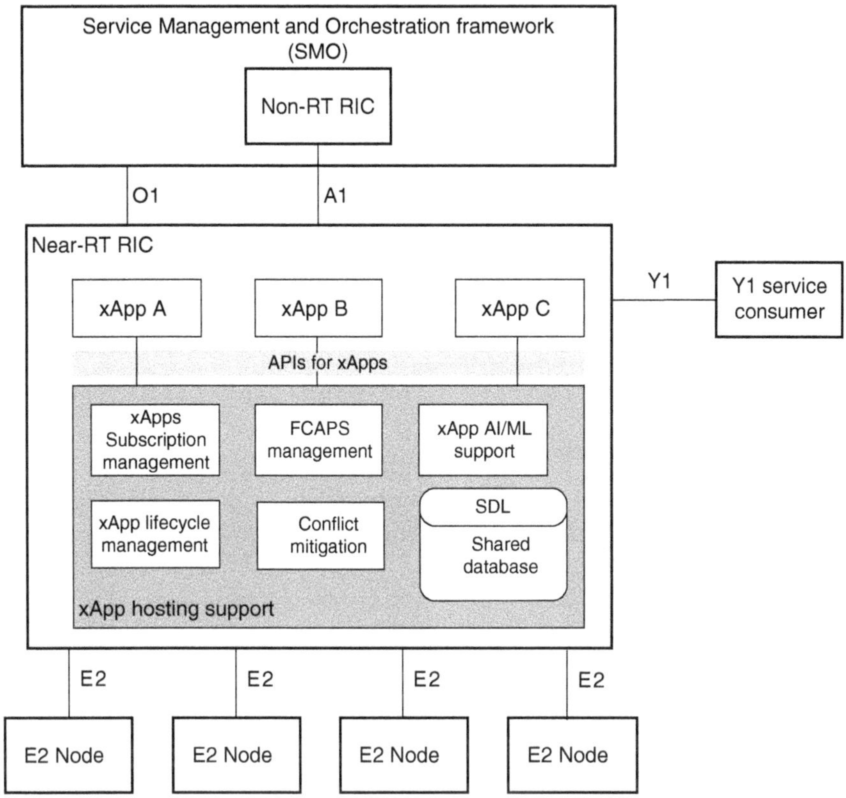

Figure 7.12 Near-RT RIC Internal Architecture. (*Source:* Vladimir Yanover.)

- Interface Termination and connectivity/routing services for the interfaces supported by the Near-RT RIC:
  - E2 interface, which enables near-real-time control of E2 Nodes, such as O-CU and O-DU. E2 connectivity service allows xApps to receive and send information via the E2 interface
  - A1 interface, which enables control of the Near-RT RIC by the Non-RT RIC. The A1 connectivity /routing service allows control of xApps via the A1 interface
  - O1 interface, which enables management of the Near-RT RIC and xApps by the SMO
  - Y1 interface used for exposure of the analytics information produced in the Near-RT RIC.
- **FCAPS Management Function**: Fault Management, Configuration Management, and Performance Management services, provided to SMO

- xApp AI/ML support, which includes facilities for setting of the AI/ML Data Pipeline, data ingestion, E2 data collection, delivery of enrichment information over A1 interface, data retrieval from the shared database, information exchange between involved xApps.
- xApps subscription management
- Conflict mitigation subsystem, which resolves conflicts between requests generated by different xApps to same E2 Node

xApps hosting support includes the xApp Repository Function and xApp LCM management by the SMO, messaging infrastructure enabling low-latency communication between Near-RT RIC internal components, and security supervision function, which provides the security scheme for xApps.

7.8.2 APIs for Support of xApps

For xApps, two implementation/deployment schemes are possible. The xApps and the Near-RT RIC Platform may be running on same HW/cloud component; in this case the interface between the xApps and the Near-RT RIC Platform can be implemented as software API supported by the corresponding SDK ("SDK Approach"). Another possibility, "Network API Approach," is that xApps and the Near-RT RIC Platform are running on separate HW/cloud components. In such case, the interface between xApps and the Near-RT RIC Platform requires use of a network protocol stack, e.g. JSON/HTTPS.

Table 7.5 describes some categories of xApp APIs.

Table 7.5 xApp APIs

API category	Description
A1-related APIs	APIs exposed by A1 Termination, used for A1 policy management and enforcement. They are also used for Enrichment Information (EI) transfer
E2-related APIs	E2-related APIs are used by xApps to access E2 Nodes via E2 interface and for E2 subscription management
Management APIs	Management APIs provide support for xApp registration at the Near-RT RIC platform. Another set of functions enables SMO management services consumers to configure the xApps and collect performance data from the xApps
Shared Data Layer (SDL) APIs	The SDL APIs allow xApps to store and retrieve data from the shared database and to subscribe to SDL notifications
Enablement APIs	The Enablement APIs allow management of access to APIs, e.g. API registration and discovery

References

1 O-RAN.WG1.OAD-R003-v11.00. O-RAN Architecture Description.
2 3GPP TS 38.410. NG-RAN; NG general aspects and principles.
3 O-RAN.WG3.UCR-R003-v05.00. O-RAN Use Cases and Requirements.
4 O-RAN.WG3.E2AP-R003-v05.00. O-RAN E2 Application Protocol (E2AP).
5 O-RAN.WG3.E2GAP-R003-v05.00. O-RAN E2 General Aspects and Principles (E2GAP).
6 O-RAN.WG3.E2SM-CCC-R003-v03.00. Near-Real-time RAN Intelligent Controller. E2 Service Model (E2SM), Cell Configuration and Control.
7 O-RAN.WG3.E2SM-KPM-R003-v04.00. Near-Real-time RAN Intelligent Controller. E2 Service Model (E2SM) KPM.
8 O-RAN.WG3.E2SM-R003-v05.00. E2 Service Model (E2SM).
9 O-RAN.WG3.E2SM-RC-R003-v05.00 Near-Real-time RAN Intelligent Controller. E2 Service Model (E2SM), RAN Control.
10 3GPP TS 38.331. Radio Resource Control (RRC) protocol specification.
11 3GPP TS 37.483. E1 Application Protocol (E1AP).
12 3GPP TS 38.413. NG application protocol (NGAP).
13 3GPP TS 38.460. NG-RAN; E1 general aspects and principles.
14 3GPP TS 38.470. NG-RAN; F1 general aspects and principles.
15 3GPP TS 38.473. F1 application protocol (F1AP).
16 3GPP TS 38.420. NG-RAN; Xn general aspects and principles.
17 3GPP TS 38.423. Xn application protocol (XnAP).
18 3GPP TS 28.552. 5G performance measurements.
19 3GPP TS 28.541. 5G Network Resource Model (NRM).
20 O-RAN.WG3.RICARCH-R003-v05.00. Near-RT RIC Architecture.

8

RAN Decomposition and Cloudification

CONTENTS

Open RAN: Technology and Ecosystem, First Edition. Vladimir Yanover.

Acronyms and Abbreviations

A1-P	A1 Policy service
A1-EI	A1 Enhancement Information service
AAL	Acceleration Abstraction Layer
AAL-LPU	Acceleration Abstraction Layer - LPU
AALI-C	Acceleration Abstraction Layer Interface-Common
AALI-C-Mgmt	Acceleration Abstraction Layer Interface-Common-Management
AALI-C-App	Acceleration Abstraction Layer Interface-Common-Application
AALI-P	Acceleration Abstraction Layer Interface-Profile
COTS	Commercial Off-The-Shelf; the term is applied to hardware, software, and services
DL	Downlink
DMS	Deployment Management Services
E2AP	E2 Application Protocol
E2SM	E2 Service Model
EI	A1 Enrichment Information
ETSI	European Telecommunication Standards Institute
FCAPS	Fault, Configuration, Accounting, Performance, Security functions
FEC	Forward Error Correction
FFT	Fast Fourie Transform
FOCOM	Federated O-Cloud Orchestration & Management
gNodeB-CU	gNodeB – Central Unit (also gNB-CU)
gNodeB-CU-CP	gNodeB – Central Unit – Control Plane (also gNB-CU-CP)
gNodeB-CU-UP	gNodeB – Central Unit – User Plane (also gNB-CU-UP)
gNodeB-DU	gNodeB – Distributed Unit (also gNB-DU)
GPU	Graphics Processing Unit
HW	Hardware
HWA	HW Accelerator
IE	Information Element
iFFT	inverse Fast Fourie Transform
IMS	Infrastructure Management Services
KPM	Key Performance Measurements
L1	Layer 1 of air interface protocol stack; also referred to as PHY Layer
L2	Layer 2 of air interface protocol stack
LAA	Licensed-Assisted Access
LBT	Listen Before Talk
LCM	Life Cycle Management
Near-RT RIC	Near-Real-Time Radio Intelligent Controller

NF	Network Function
NFO	Network Function Orchestration
NFV	Network Functions Virtualization
NIC	Network Interface Card
Non-RT RIC	Non-Real-Time Radio Intelligent Controller
NS	Network Service: the term defined in ETSI GR NFV 003 [1] as "composition of Network Function(s) and/or Network Service(s), defined by its functional and behavioral specification." This terminology is used also in 3GPP 28-series of specifications. NOTE. This definition is essentially different from commonly used term "Network Service" meaning a service provided to subscribers
O-Cloud	A cloud platform, supporting hosting of O-RAN network functions
O-CU	O-RAN Central Unit
O-CU-CP	O-RAN Central Unit – Control Plane
O-CU-UP	O-RAN Central Unit – User Plane
O-DU	O-RAN Distributed Unit
O-RU	O-RAN Radio Unit
OFH	Open FrontHaul interface defined in O-RAN (also Open FH)
PNF	Physical Network Function
PRB	Physical Resource Block
rApp	Third-party application hosted by the Non-RT RIC
UDP	User Datagram Protocol
UL	Uplink
VNF	Virtualized Network Function
xApp	Third-party application hosted by the Near-RT RIC

8.1 Introduction

One of fundamental innovations proposed by O-RAN ALLIANCE is further RAN decomposition, also called "RAN split" or sometimes "RAN disaggregation."

For 5G, 3GPP defined RAN split into gNodeB-CU (Central Unit) and gNodeB-DU (Distributed Unit); see Figure 6.1-1 in 3GPP TS 38.401 [2]. The gNodeB-CU can be further split into gNodeB-CU-CP (Control Plane) and gNodeB-CU-UP (User Plane). The 3GPP RAN architecture study, in the TR 38.801 [3], identified several possible options of CU/DU functional split; at this point only one of them, Option 2, is addressed in 3GPP.

O-RAN architecture uses Option 2 split between the gNodeB-CU and gNodeB-DU but added split of gNodeB-DU into two components: O-DU (O-RAN Distributed Unit) and O-RAN Radio Unit (O-RU). The functional split between O-DU and O-RU follows Option 7-2 specified in 3GPP TR 38.801 [3].

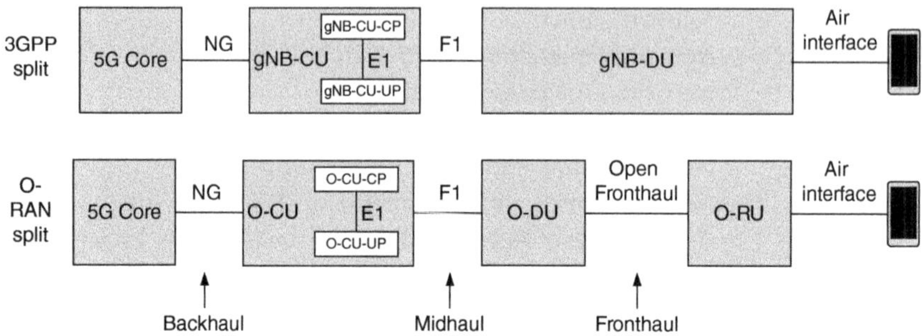

Figure 8.1 RAN decomposition in O-RAN. (*Source:* Vladimir Yanover.)

Figure 8.1 shows mapping between 3GPP RAN split and O-RAN split. The O-RAN split scheme includes NG and F1 interfaces defined in 3GPP, with addition of a new interface, Open Fronthaul (OFH), between newly defined components of the gNodeB-DU: O-DU and O-RU. There is also some functionality extension caused by introduction of Near-Real-Time Radio Intelligent Controller (Near-RT RIC) and the corresponding interface E2; see Chapter 7 for this part. For example, O-CU functionality includes complete gNodeB-CU functionality with addition of the E2 interface support.

Cloudification of the O-CU is comparatively straightforward. For the O-DU, it is more challenging, because the O-DU implements computationally intensive Layer 2 and a part of Layer 1 of the air interface, which may require use of powerful RAN HW accelerators (Section 8.5).

In this chapter, Section 8.2 describes the **OFH interface,** based on specification O-RAN.WG4.CUS.0-R003-v14.00 [4], in view of supported O-DU/O-RU functional split. In most scenarios, O-RU is a Physical Network Function (PNF) placed at a remote cell site location. The cloud hosting the O-DU should provide the O-DU with OFH connectivity. Another requirement to the cloud is to provide the O-DU with HW acceleration capabilities for support of the Layer 1 and 2 processing and of the OFH interface.

Section 8.3 describes **O-Cloud Platform** (O-RAN.WG6.CADS-V06.00 [5], O-RAN.WG6.ORCH-USE-CASES-R003-v09.00 [6]), and **O-Cloud management and orchestration** (O-RAN.WG6.O2DMS-INTERFACE-ETSI-NFV-PROFILE-R003-v06.00 [7] O-O-RAN.WG6.O2DMS-INTERFACE-K8S-PROFILE-R003-v04.00 [8], O-RAN.WG6.O2-GA&P-R003-V06.00 [9], O-RAN.WG6.O2IMS-INTERFACE-R003-V05.00 [10])

Section 8.4 outlines **Cloud deployment scenarios**, which describe placement of the O-RAN VNFs and PNFs in the O-Cloud components (O-RAN.WG6.CADS-V06.00 [5]).

Section 8.5 describes **HW acceleration** technology based on O-RAN specifications O-RAN.WG6.AAL-Common-API-R003-v06.00 [11], O-RAN.WG6.AAL-GAnP-R003-v08.00 [12], O-RAN.WG6.AAL-HI_PHY-R003-V05.00 [13]. The ETSI

NFV specification ETSI GR NFV-IFA 046 [14] can be useful for analysis of how the NFV framework can support virtualized RAN.

8.2 Open Fronthaul (OFH) Interface

8.2.1 Introduction

This section describes O-DU/O-RU functional split and the OFH interface.

There is a tradeoff for functional split between O-DU and O-RU. Split at a lower layer requires higher user plane bit rate on the OFH interface, which may be not available or overly expensive in some deployments. On the other hand, lower layer split means less functionalities in the O-RU, which generally results in lower complexity/cost and lower power consumption. O-RAN ALLIANCE selected intra-PHY split 7-2, i.e. at comparatively low layer. Such solution requires OFH to carry IQ level user data, therefore to support high data rates at the OFH User Plane, which usually is achieved by using optical fiber links.

8.2.2 O-DU/O-RU Functional Split

The specification O-RAN.WG4.CUS.0-R003-v14.00 [4] describes two slightly different split options under Split 7-2, targeting two O-RU categories, A and B. The difference between A and B is that O-RU category B contains Precoding block, while O-RU category A does not. Accordingly, in case of category B, the OFH interface contains a part related to precoding.

Figure 8.2 describes the functional split over the Open Fronthaul interface, between the O-DU and O-RU for both Categories A and B. More details can be found in O-RAN.WG4.CUS.0-R003-v14.00 [4], Section 4.2.

The design of the OFH protocol follows the provisions of the selected functional split. The User Plane of the OFH protocol carries IQ samples, while the Control

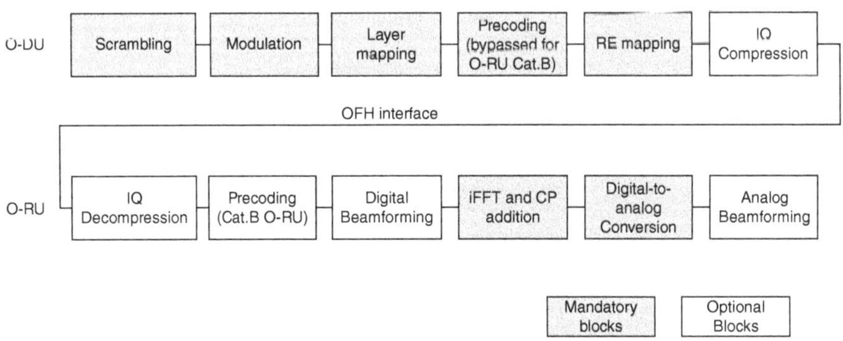

Figure 8.2 OFH Split; Category A and Category B O-RUs. (*Source:* Vladimir Yanover.)

Plane protocol allows configuration of the O-RU functions and the control over OFH scheduling and beamforming.

8.2.3 Data Flow Over OFH Interface

The specification O-RAN.WG4.CUS.0-R003-v14.00 [4] defines OFH Control Plane, User Plane, and Sync Plane (Table 8.1). Note that the OFH User Plane ("U-Plane") includes also the data carried via the Control Plane of the Air Interface. The OFH User Plane carries the IQ sample data between O-DU and O-RU; in the User Plane the following flows can be identified (definition of 5G channels PDSCH, PDCCH, PUSCH, PUCCH, and PRACH can be found in 3GPP TS 38.300):

- Flows of IQ sample data on DL, including PDSCH, PDCCH
- Flows of IQ sample data on UL, including PUSCH, PUCCH, and PRACH

The OFH Control Plane ("C-Plane") refers to real-time OFH control information transferred from the O-DU to the O-RU. This information includes scheduling commands and beamforming commands for DL and UL, and setting the corresponding parameters with granularity down to individual Resource Element (RE) per beam per slot.

The OFH Synchronization Plane (S-Plane) refers to Timing and Synchronization data transferred between the O-DU and O-RU and processed by the synchronization controller.

The M-Plane for the OFH interface provides for management capabilities; it is specified in O-RAN.WG4.MP.0-R003-V14.00 [15].

Figure 8.3 shows the role and timing of the scheduling/beamforming commands. The C-Plane messages contain control information, which guides processing of user data in the O-RU. If transmitted to the same direction, the control information can be combined into a single C-Plane message. For example, C-Plane information of PUCCH and PUSCH data may be combined into a single C-Plane message.

Table 8.1 Open Fronthaul data flows

Data flow	**Direction**	**Notes**
OFH Control Plane	O-DU to O-RU	Scheduling information, Beamforming commands
OFH User Plane, DL	O-DU to O-RU	DL IQ Data: PDSCH, PDCCH, . . .
OFH User Plane, UL	O-RU to O-DU	UL IQ Data: PUSCH, PUCCH, . . .
OFH User Plane, UL	O-RU to O-DU	UL IQ Data: PRACH
OFH Synchronization Plane	O-DU to O-RU	Timing and Synchronization information

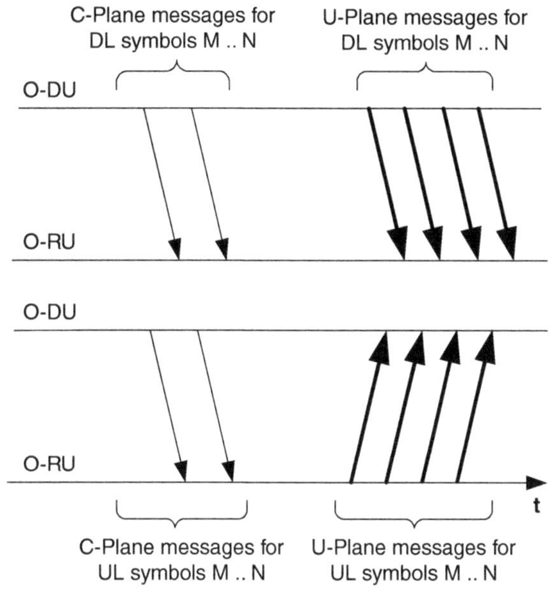

Figure 8.3 OFH Control Plane and User Plane. (*Source:* Vladimir Yanover.)

8.2.4 OFH Protocol Architecture

8.2.4.1 Transport: eCPRI or RoE

Figure 8.4 shows the protocol stack used for exchange of C-Plane and U-Plane messages between the O-DU and O-RU. The stack includes the following layers:

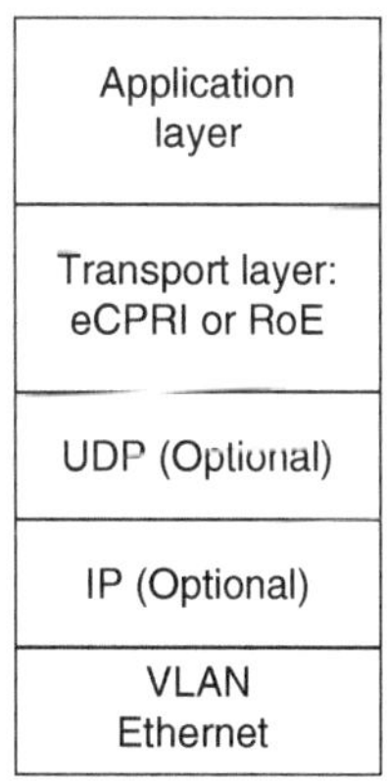

Figure 8.4 C-Plane and U-Plane protocol stack. (*Source:* Vladimir Yanover.)

- Application Layer
  Used by both C-Plane and U-Plane (Section 8.2.4.2).
- Transport Layer
  Either eCPRI protocol [16] or IEEE 1914.3 protocol (also known as RoE, [17]) is used. In case of eCPRI, indication of the transport payload type is done by ecpriMessage value: 0x0 for U-Plane data and 0x2 for C-Plane data. In case of RoE, it is indicated by the RoEsubType field.
- UDP, IP Layers
  The Transport layer messages can optionally be encapsulated into UDP/IP packets.
- Ethernet, Layers 1 and 2
  The underlying Ethernet layer enables transfer of eCPRI or RoE or IP messages as the Payload field of the 802.3 Ethernet II/DIX frame with Type interpretation of the Type/Length field. The Payload field can contain up to

9000 bytes. It is assumed that the MAC address resolution for O-DU and O-RU is performed via an M-Plane procedure described in O-RAN.WG4.MP.0-R003-V14.00 [15]. The Ethernet frame contains VLAN Tag field defined in IEEE 802.1Q with 3-bit Priority Code Point (PCP) sub-field used for QoS differentiation; VLAN tagging is M-Plane configurable.

For transfer of the C-Plane and U-Plane information, O-RAN specifications define an application layer on top of Transport layer (Figure 8.4).

The eCPRI message consists of the Common Header and eCPRI Payload [16]. Figure 8.5 shows use of the eCPRI format in the Open Fronthaul interface. First 4 bytes contain eCPRI Common Header. The type of the eCPRI Payload content is indicated by the eCPRI Message Type filed: 0 for U-Plane, 2 for C-Plane. The CI field indicates concatenation of the protocol messages. The eCPRI Payload begins with two OFH-specific fields, each of which consists of 2 bytes: the ecpriRtcid/ecpriPcid field identifies the data flow within the C-Plane/U-Plane; the ecpriSeqid is a message sequential number.

8.2.4.2 C-Plane and U-Plane Application Layer

Within the application layer messages, the SectionId value links between the C-Plane message and related U-Plane messages. The SectionType defines the format of the C-Plane message and identifies the L1 characteristics for transmission/reception of the corresponding data, encoded in the corresponding U-Plane messages with same SectionId.

There are eight types of sections linked to the categories of C-Plane data. For example, C-Plane Section Type 3 can be used for transfer of the information of mixed-numerology channels. Such section identifies the time–frequency area to which the section is applicable in terms of frame (frameId), subframe (subframeId), slot (slotID), PRBs (startPrbc,

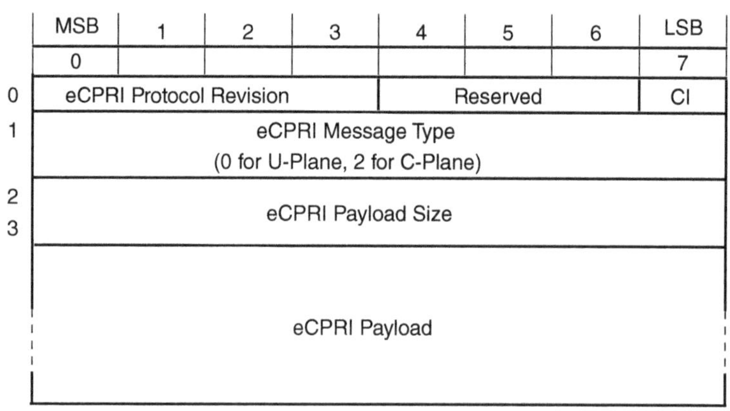

Figure 8.5 eCPRI message format in the OFH interface.

numPrbc), symbol intervals (startSymbolId, numSymbol). Further, the section links that area to the FFT/iFFT size and subcarrier spacing (frameStructure). To control beamforming in the O-RU, the section identifies the beam pattern (beamId).

Other examples of section types: Type 1 for C-Plane data, such as scheduling and beamforming commands, Type 6 for sending UE-specific channel information from the O-DU to the O-RU, etc. The format definitions for all types of sections can be found in O-RAN.WG4.CUS.0-R003-v14.00 [4], Sections 7.4 and 7.5.

For U-Plane, similar definitions are provided in [4], Section 8. In the corresponding protocol sections, the IQ-encoded data appears in the format of iSample, qSample fields (1–16 bits).

8.2.4.3 S-Plane

There are two mechanisms for frequency and time synchronization between the O-DU and O-RU connected by Ethernet link. One mechanism is using Synchronous Ethernet (SyncE) ITU-T standard, with synchronization based on clock signals used by the Ethernet physical layer. SyncE was standardized by the ITU-T as three ITU-T Recommendations: G.8261, G.8262, and G.8264. Another mechanism is using IEEE 1588 Precision Time Protocol (PTP) [18] with PTP messages transferred directly over Layer 2 Ethernet, per ITU-T Recommendation G.8275.1 [19]. More details on the frequency and time synchronization can be found in Chapter 11 of O-RAN.WG4.CUS.0-R003-v14.00 [4].

8.3 O-Cloud Platform

8.3.1 O-Cloud Architecture

O-Cloud is a cloud platform hosting O-RAN network functions; it exposes defined in O-RAN APIs enabling LCM of O-RAN network functions; see O-RAN.WG6.CADS-V06.00 [5].

In line with modern virtualization/cloudification concepts, the Cloud Platform hardware includes Commercial-Off-The-Shelf (COTS) compute, networking, and storage components; it may also include hardware accelerators. Use of accelerators is justified in the case when the required combination of cost and performance is beyond available COTS compute capabilities. Use of accelerators is particularly important in the O-DU for execution of computationally intensive RAN Layer 2 and 1 functions. On the other hand, for Near-RT RIC and O-CU-CP, general-purpose COTS servers usually provide sufficient level of performance.

The Cloud Platform software contains an orchestration layer, such as Kubernetes, and exposes well-defined APIs, which enable Life Cycle Management (LCM) of the network functions and network services hosted by the platform.

Open APIs allow decoupling the platform's software from the platform's hardware, so they can be produced by different vendors.

Table 8.2 describes O-Cloud Platform containment relationships.

Figure 8.6 depicts the structure of the O-Cloud Platform. The figure shows allocation of compute resources to the Nodes; other types of resources allocated to the Nodes are not shown in the figure.

The figure also shows O2ims and O2dms interfaces, coming from the SMO (O-RAN.WG6.O2-GA&P-R003-V06.00 [9]). These interfaces are used by the SMO for control of the following functions:

- Management of the cloud infrastructure: Infrastructure Management Services (IMS)
- Life Cycle Management (LCM) of VNFs and NSs running on the cloud infrastructure: Deployment Management Services (DMS)

NOTE. "NS" stands for Network Service as defined in ETSI GR NFV 003 [1]; it's essentially different from the definition of "Network Service" in 3GPP. In ETSI

Table 8.2 Relationships between O-Cloud Platform components

O-Cloud Platform component	Contains components
O-Cloud	• Multiple O-Cloud Sites, interconnected with corresponding network technology. • An O-Cloud Node Cluster
O-Cloud Site	A set of pooled O-Cloud Resources: • compute • storage • HW accelerators NOTE. A resource pool can be distributed across multiple O-Cloud Sites • Network Fabrics used for interconnecting the O-Cloud Resources within the Site; usually with Layer 2 switching
O-Cloud Node Cluster	O-Cloud Nodes, connected by Node Cluster Networks, that work in coordination with each other. The Nodes can be optionally aggregated in Node Groups. An O-Cloud Node is provisioned into the O-Cloud Node Cluster by the Infrastructure Management System (IMS) NOTE. An O-Cloud Node could be also set as a stand-alone O-Cloud Node
O-Cloud Node	O-Cloud Nodes can include CPUs, memory and storage resources, HW Accelerators, etc.

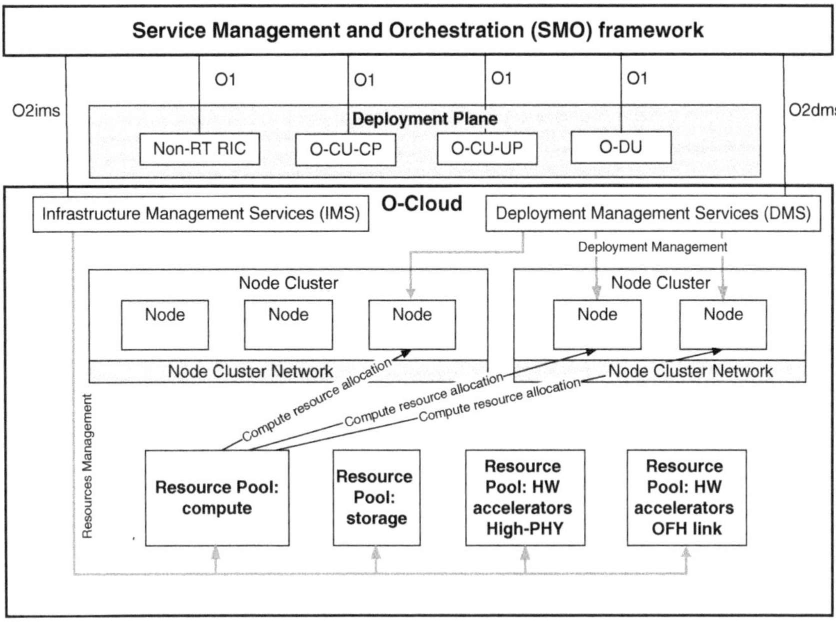

Figure 8.6 Structure of O-Cloud Platform. (*Source:* Vladimir Yanover.)

NFV, "Network Service" is a composition of connected Network Functions, defined by its functional and behavioral specification.

The Deployment Plane in the figure shows examples of O-RAN functions deployed on the O-Cloud.

8.3.2 Infrastructure Management Services

This section outlines the subset O2ims of the O2 interface, responsible for deployment and management of the cloud infrastructure, according to the specification O-RAN.WG6.O2IMS-INTERFACE-R003-V05.00 [10]. This subset includes the following services exposed by the O-Cloud to the SMO:

- **Inventory Services**
  These services allow querying availability of resources in the cloud: physical or logical CPUs, GPUs, Memory blocks, Storages, NICs, HW Accelerators, etc. The service also allows querying the O-Cloud about Resource Pools and querying the Deployment Managers installed in the system. Additionally, SMO can use the Inventory Change Subscription service to subscribe to notifications on the changes in the inventory.

- **Monitoring Services**
  These services enable continuous real-time collection of O-Cloud faults and performance metrics (the latter is under development). The collected data sometimes is called "telemetry."
- **Provisioning Services**
  These services allow configuration of the O-Cloud infrastructure (under development in O-RAN)
- **Software Management Services**
  These services enable the process of software deployment and update (under development in O-RAN)
- **Infrastructure Life Cycle Management Services**
  Includes support for notification of O-Cloud availability. This notification is generated in the process of the O-Cloud establishment and contains callback URI for the consumer's registration. Planned further development includes notification of the LCM events in the O-Cloud, such as scaling.

8.3.3 Deployment Management Services

This section describes the subset O2dms of O2 interface, responsible for Life Cycle Management (LCM) of VNFs and NSs running on the cloud infrastructure. For that, O-RAN defined two architecture profiles: one is specified in O-RAN.WG6. O2DMS-INTERFACE-ETSI-NFV-PROFILE-R003-v06.00 [7]; it is based on ETSI NFV specifications. Another architecture option is specified in O-RAN.WG6. O2DMS-INTERFACE-K8S-PROFILE-R003-v04.00 [8]; it is based on Kubernetes, also known as K8s, an open-source system for automation of deployment, scaling, and management of containerized applications.

For the ETSI NFV option, Figure 8.7 shows simplified NFV-MANO architectural framework (GS NFV-MAN 001 [20]) with additional information on mapping to SMO and O-Cloud architecture. In this option ("O2-D demarcation #1" in O-RAN.WG6.O2DMS-INTERFACE-ETSI-NFV-PROFILE-R003-v06.00 [7]), the NFV Orchestrator (NFVO) and Element Management (EM) components are placed in the SMO; the VNF Manager (VNFM), NFV Infrastructure (NFVI), and Virtualized Infrastructure Manager (VIM) are placed in the O-Cloud. Management interface between the EM and Virtualized Network Function (VNF) is mapped to the O-RAN O1 interface. The Or-Vnfm interface between the NFVO and VNFM is mapped to the O-RAN O2dms interface.

The O-RAN Network Function (NF) can be mapped onto ETSI VNF or ETSI NFV Network Service (ETSI NFV NS). The latter is in order when the O-RAN NF contains several VNF instances. It should be noted that the ETSI NFV NS composition is not visible via the O2dms; information of the ETSI NFV NS composition resides in the SMO functions (NFVO), that are managing the O-Cloud via the O2dms.

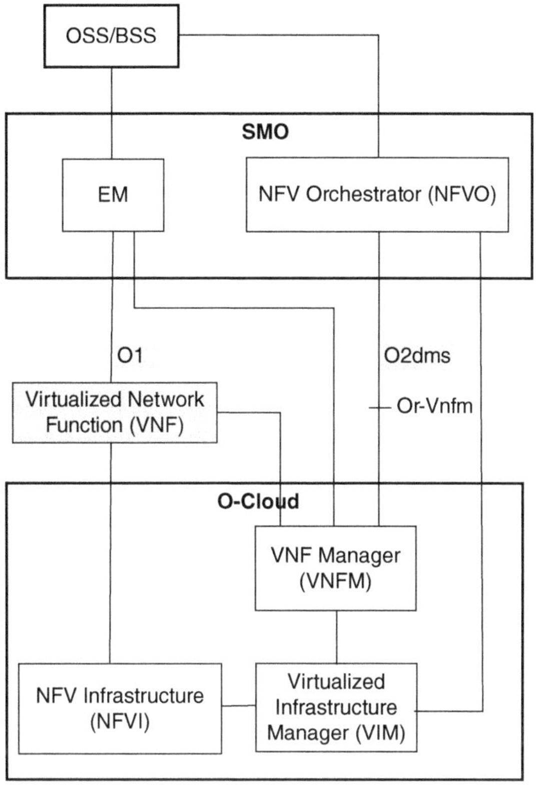

Figure 8.7 Mapping of SMO and O-Cloud to ETSI NFV architecture. (*Source:* Vladimir Yanover.)

Additional details on mapping between O-RAN components and ETSI NFV components:

- O-RAN NF Deployment Unit is mapped onto ETSI NFV VNFC
- O-RAN NF Software Package is mapped onto ETSI NFV VNF Package
- O-RAN NF Deployment Descriptor is mapped onto ETSI NFV VNFD

8.4 Cloud Deployment Scenarios

8.4.1 Classification of Deployment Scenarios

O-RAN specification O-RAN.WG6.CADS-V06.00 [5] contains classification of major cloud deployment scenarios; see Table 8.3. This classification is based on placement of O-RAN components, such as O-DU, in different O-Cloud platforms (O-RAN. WG6.CADS-V06.00 [5], Section 6.3). One O-Cloud platform, e.g. Edge Cloud, may contain multiple O-Cloud sites in different geo locations. The classification of the

Table 8.3 Major deployment scenarios, identified by O-RAN

Scenario	Near-RT RIC	O-CU	O-DU	O-RU
A	O-Cloud			Physical NF
B	----RegionalCloud----	----- Edge Cloud -----		Physical NF
C	----- Regional Cloud -----		Edge Cloud	Physical NF
C1, C2	----- Regional Cloud -----		Edge Cloud	Physical NF
D	----- Regional Cloud -----		Physical NF	Physical NF
E, E1	----- Regional Cloud -----		Cell site, O-Cloud and Physical NF	
F	----- Regional Cloud -----		Edge Cloud	Edge Cloud

scenarios still holds when some of O-RAN components are combined (integrated) in a single VNF or PNF, in which case interfaces between combined components may not be exposed. For example, in Scenario A, O-CU and O-DU can be combined in a single NF, with F1 interface exposed or not. It should be noted that in the case of integrated O-CU and O-DU, the management information models of O-CU and O-DU are still exposed to the management system. The table does not show split of O-CU into O-CU-CP and O-CU-UP, which means they either are both PNFs or are both implemented in VNFs placed in the same O-Cloud component.

Table 8.3 identifies the following O-Cloud locations:

- **Regional Cloud**
  It is a cloud platform hosting RAN VNFs serving plurality of Cell Sites in multiple Edge Clouds. The Regional Cloud therefore hosts highly centralized functions. In particular, in all scenarios the Near-RT RIC is hosted by the Regional Cloud. Selection of physical location of the Regional Cloud usually is influenced by network latency requirements.
- **Edge Cloud**
  It is a cloud platform hosting RAN VNFs serving plurality of Cell Sites. Typically, there are stringent network latency requirements, which implies that the physical location of the Edge Cloud should be close enough to the Cell Sites location to provide low latency, e.g. between O-DU and O-RUs.
- **Cell Site**
  The Cell Site usually supports multiple sectors and includes multiple O-RUs. In most deployment scenarios Cell Site contains O-RUs, which are PNFs. But in the scenarios E and F some part of the Cell Site equipment can be cloudified.

Figure 8.8 shows recommendations developed by O-RAN ALLIANCE for user plane latency targets, associated with physical distances between different O-Cloud components [5]. For the air interface, the latency targets are not O-RAN specific.

Different user services may have different user plane latency requirements, in which case the network operator may choose different placements for the

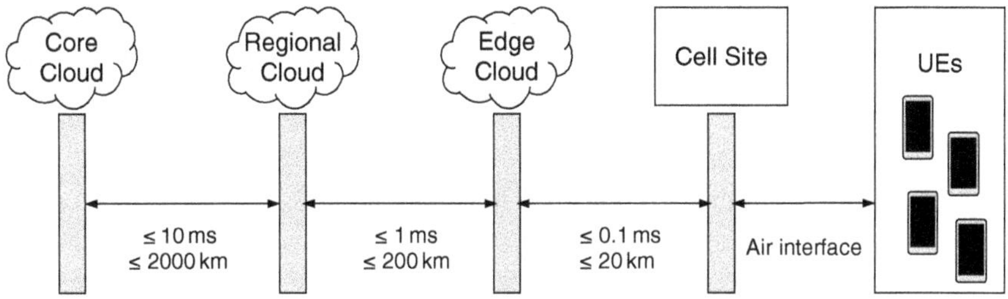

Figure 8.8 User plane latency targets. (*Source:* Vladimir Yanover.)

corresponding User Plane components. Figure 8.9 shows an example of deployment, supporting two network slices of the eMBB and URLLC type. For the eMBB slice, the O-CU-UP can be placed in the Regional Cloud, with O-DU placed in the Edge Cloud, which corresponds to Scenario C. The URLLC slice has more stringent latency requirements, so the operator may choose placing the O-CU-UP in the Edge Cloud, which can be considered a variant of Scenario B. The figure also shows placement of the UPF 5G Core function. Figure 8.9 shows separate O-CU-UP and O-DU instances for different slices, but specific implementation of O-CU-CP or O-DU may provide for simultaneous support of multiple slices with different types. The figure shows two different UEs served via different slices.

Table 8.4 contains O-RAN recommendations for cardinality in some deployment scenarios.

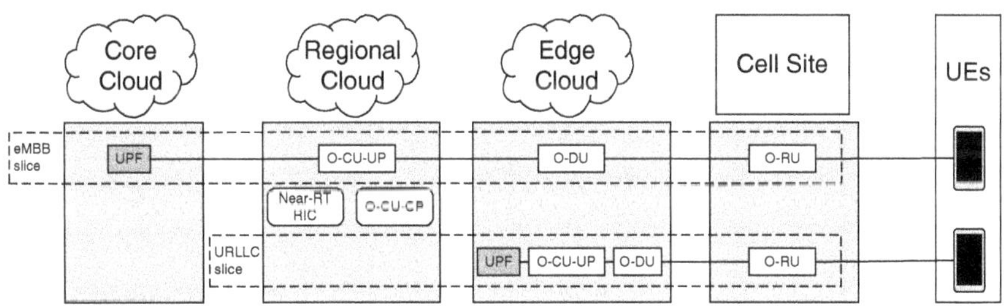

Figure 8.9 Example of service-type specific placement of RAN components. (*Source:* Vladimir Yanover.)

Table 8.4 Cardinality in some deployment scenarios

Scenario	Number L of O-CUs per Near-RT RIC	Number M of O-DUs per O-CU	Number N of O-RUs per O-DU
Scenario B	100+	1	1–64
Scenario C	1	100+	32–64
Scenario C.1	1	320	100

8.4.2 Overview of Selected Deployment Scenarios

In all scenarios, the cloud platform, hosting O-DUs, should be able to supply hardware acceleration for Layer 1 and 2 computationally intensive procedures. See Section 8.5 for more details.

In Scenario A (Figure 8.10), the Near-RT RIC, O-CU, and O-DU are implemented as VNFs running on the same cloud component, Edge Cloud platform (Figure 8.10). The interfaces E2 and F1 are internal to the Edge Cloud platform.

This scenario requires high fronthaul capacity to carry User Plane traffic between O-DU in Edge Cloud and O-RU in the Cell Site. Combined with low latency requirements on the fronthaul link, it more likely applies to dense urban areas.

In Scenario C (Figure 8.11), the Near-RT RIC and O-CU are hosted by the Regional Cloud platform, while the O-DU is hosted by the Edge Cloud platform. In this scenario, the interfaces F1 and E2 to O-DU go via the transport network; multiple O-RUs are connected to the pooled O-DU function. The scenario is tuned for support of a dense urban deployment with large number of radio cells, some of which may have a small coverage area (e.g. in mmWave).

Scenario C.2 is derived from Scenario C, to address the case of network slicing with different latency targeted by different slices. Figure 8.12 shows connection

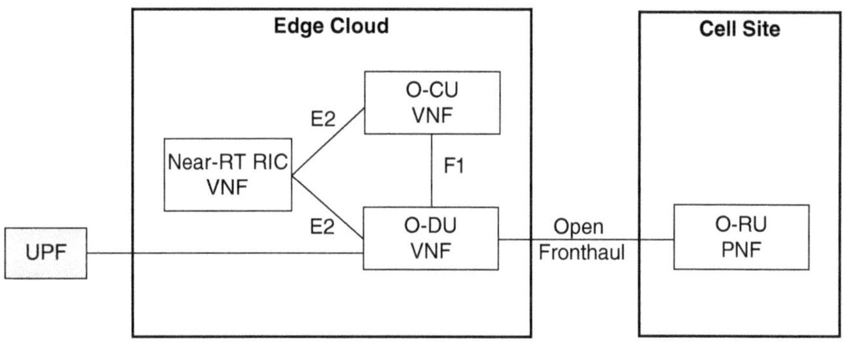

Figure 8.10 Deployment scenario A. (*Source:* Vladimir Yanover.)

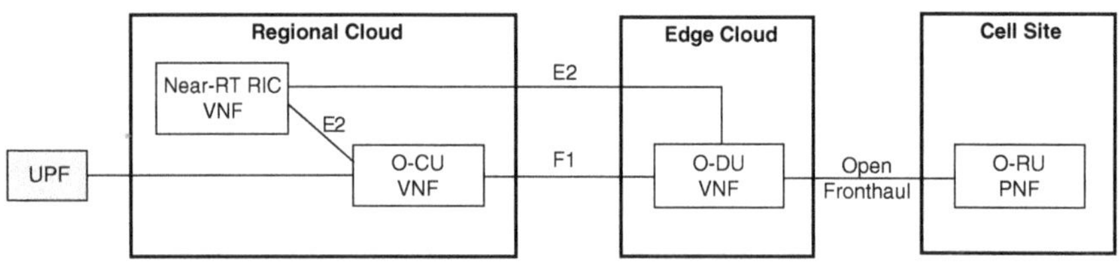

Figure 8.11 Deployment Scenario C. (*Source:* Vladimir Yanover.)

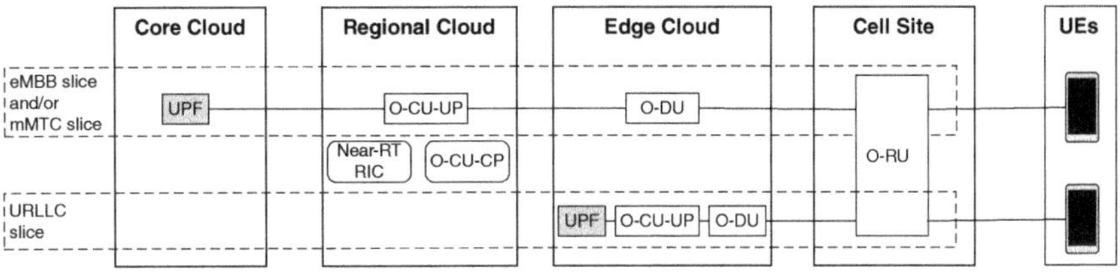

Figure 8.12 Deployment Scenario C.2. (*Source:* Vladimir Yanover.)

from the User Plane Function (UPF) terminated in the Regional Cloud for eMBB and mMTC network slices. The URLLC slice is using both the UPF, and the O-CU placed in the Edge Cloud. Different slices may use different O-DUs located in the Edge Cloud.

Scenario C.2 allows O-RU sharing between multiple network operators, with granularity of network slice.

Scenario C.1 is similar to Scenario C.2, with the only difference that it is using an integrated O-DU in the Edge Cloud, serving multiple slices. For the scenario C.1, the specification O-RAN.WG6.CADS-v06.00 [5] recommends 100 μs (20 km distance) as the target for the latency between O-DU and O-RU. The recommended target for the latency between the O-CU and O-DU in case of mMTC, eMBB, and the Control Plane of the URLCC, is 625 μs (125 km). For the URLCC User Plane, the target is 100 μs (20 km).

8.5 Hardware Acceleration

8.5.1 Introduction

As mentioned in 8.1, an O-Cloud node may include hardware accelerators to optimize performance of cloudified network functions (VNFs), especially the O-DUc. In many cases, accelerators are implemented as expansion cards installed in the corresponding COTS servers. Accelerator implementation may include ASICs, FPGAs, Graphic Processing Units (GPUs). Examples of accelerated RAN functions can be found in Section 8.5.3, which includes the Layer 1 processing blocks for O-DU; see also Figure 7.1-1 in O-RAN.WG6.AAL-GAnP-R003-v08.00 [12].

The unique approach taken by O-RAN is to enable portability of the RAN software across hardware platforms, using O-RAN compliant HW accelerators. For this purpose, O-RAN ALLIANCE defined the HW acceleration architecture and specified the corresponding interfaces (APIs).

For scheduling of accelerator actions, O-RAN selected a burst-oriented approach, where the unit is a burst of operations performed by the acceleration

HW. Such burst is enqueued in one of the input queues at the corresponding processing unit. After processing, the burst is dequeued.

O-RAN specification O-RAN.WG6.AAL-GAnP-R003-v08.00 [12] defines basic terminology for HW accelerator architecture (Figure 8.13):

- **AAL** means Acceleration Abstraction Layer
- **AAL Application (AAL-App)** is a workload (such as Layer 1 processing); the AAL-App can offload some functions to HW Accelerator
- **AAL Profile** is an abstraction, which defines one or several accelerated functions (AAL Profile instances) executed by the HW accelerator for the benefit of an AAL Application
- **AAL Profile Instance** is a part of AAL-App workload offloaded to an AAL-LPU, according to specific AAL Profile
- **AAL Logical Processing Unit (AAL-LPU)** is an abstraction of certain portion of HW Accelerator resources, which is allocated to one or more of AAL Profile Instances
- **AAL Queue** is exposed by the AAL Profile Instance associated with an AAL-LPU. The queue contains bursts of operations to be performed by the AAL-LPU. A burst can be added to the queue using the Enqueue API. The processed burst is removed from the queue using the Dequeue API. Split of the workflow into multiple queues may be applied to separate use of specific resources (compute, I/O) or to prioritize between queues.

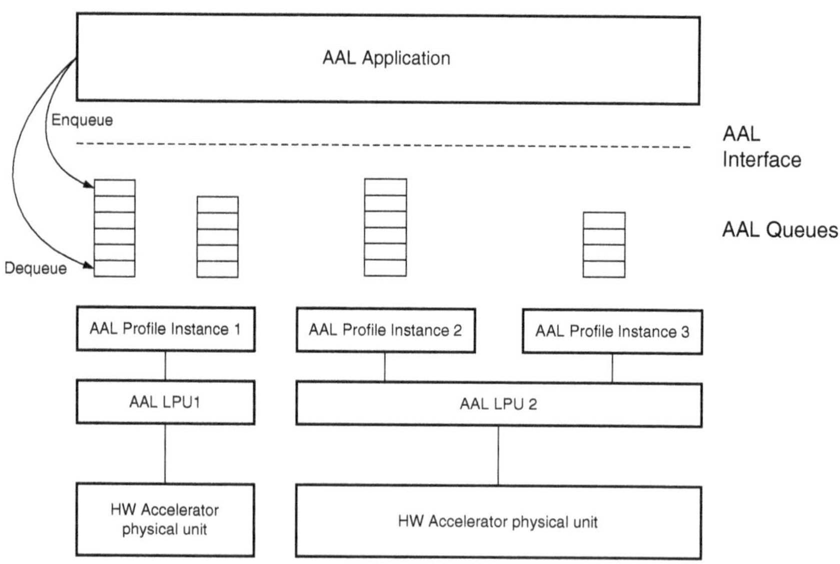

Figure 8.13 HW Acceleration architecture. (*Source:* Vladimir Yanover.)

There are two competing hardware accelerator deployment models: look-aside processing and inline processing. Historically these two models were defined in ETSI specification ETSI GS NFV-IFA 002 [21].

Figure 8.14 shows look-aside model where the host CPU invokes a HW accelerator and transfers to it a data block to be processed. After processing, the host CPU receives the processing result and initiates the next processing step. In this model, overall processing time includes the time of resulting data block transfer back to the CPU, the CPU reaction time (with associated jitter) and transfer of the data block to the next processing step.

As different from look-aside architecture, in the inline accelerator architecture the accelerator combines data processing with transfer of the resulting data to the next processing step (Figure 8.15). Therefore, for inline acceleration model, the

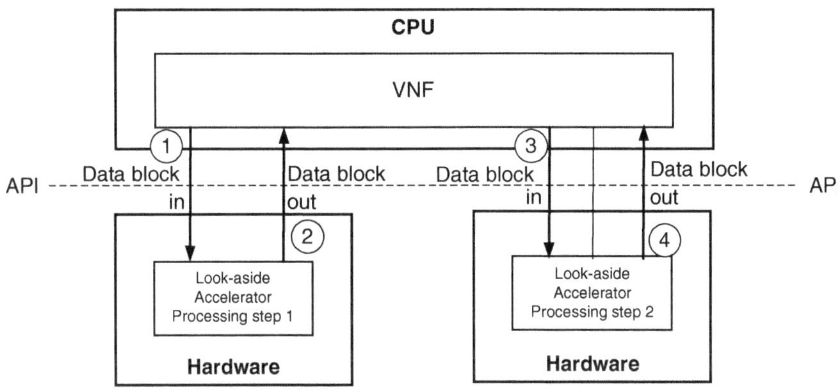

Figure 8.14 AAL look-aside acceleration model. (*Source:* Vladimir Yanover.)

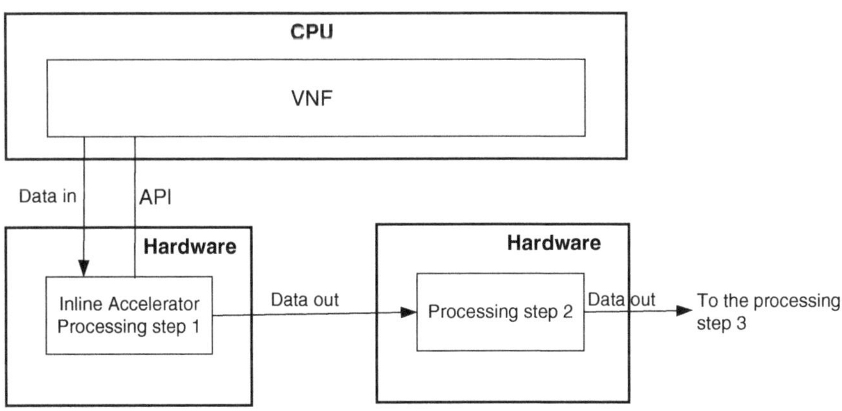

Figure 8.15 AAL inline acceleration model. (*Source:* Vladimir Yanover.)

processing time does not include transfer of resulting data block back to the CPU, the CPU reaction time, and transfer of the resulting data block to the next processing step. This is why inline architecture in general provides for significantly lower processing latency.

Software implementation can be based on one of the following models defined in ETSI GS NFV-IFA 002 [21]:

- **PCI pass-through model with SR-IOV feature**

In this model, the VNF must contain a driver for every HW accelerator used by the VNF.

- **Abstracted model deployment type**

This model is using a functional abstraction of the HW accelerator; the corresponding interface is exposed by the adaptation layer in the NFV infrastructure. The VNF supports only the interface to the adaptation layer, which is independent from specific implementation of the HW accelerator.

8.5.2 Acceleration Abstraction Layer (AAL)

Acceleration Abstraction Layer (AAL) interface was defined in O-RAN ALLIANCE to enable multi-vendor support for hardware accelerators. HW accelerators supporting the standard AAL Interface and VNFs supporting the same interface may come from different vendors. The AAL Interface API supports both look-aside and inline models.

Figure 8.16 outlines hardware acceleration architecture with Acceleration Abstraction Layer (AAL), which includes:

- **AAL Application** running in the Process Address Space, a part of the O-RAN network function.
- **AAL implementation**: HW Accelerator together with accompanying software, such as a kernel space driver or library or user space driver.

The interface between the AAP Application and the library/driver software is called AAL Interface; it is an open interface standardized in O-RAN ALLIANCE for several types of HW Accelerators. One AAL Application can use multiple HW accelerator instances (AAL Logical Processing Units or AAL-LPUs).

8.5.3 AAL Profile

According to O-RAN.WG6.AAL-GAnP-R003-v08.00 [12], an AAL Profile of a Hardware Accelerator specifies a set of functions that can be offloaded to the

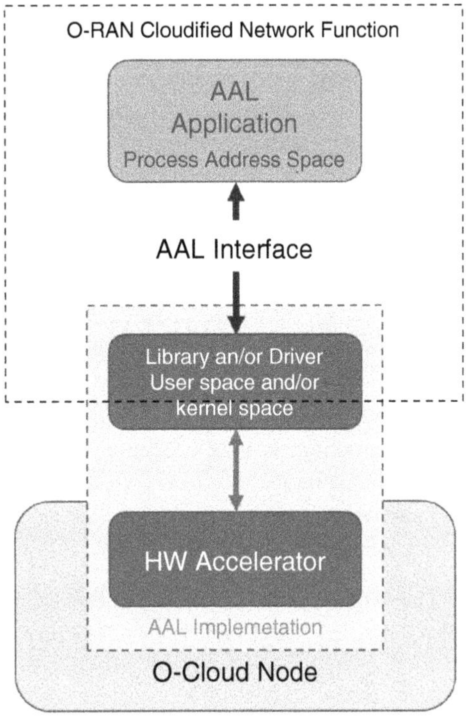

Figure 8.16 AAL Specification Scope. (*Source:* Reproduced with permission of O-RAN ALLIANCE e.V.)

accelerator. The profiles are categorized by types of AAL Applications serviced by the Accelerator.

The specification O-RAN.WG6.AAL-GAnP-R003-v08.00 [12] defines several sets of AAL Profiles, with names following a special pattern. For example, the family of O-DU acceleration profiles is based on the model of the HIGH-PHY stack; the names of the profiles follow the pattern AAL_*_HIGH-PHY. The model for downlink "O-DU PHY processing blocks for 5G NR Downlink" is depicted in the Figure 5-1 in [12]. For example, the profile AAL_PDSCH_HIGH-PHY includes the processing of PDSCH Transport Blocks (TBs) and associated Demodulation Reference Signal (DM-RS):

- **PDSCH-TBs**
  - TB CRC attachment
  - Code Blocks (CBs) segmentation and CRC attachment
  - LDPC encoding
  - Rate matching
  - CB concatenation
  - Scrambling

- Modulation
- Layer mapping
- Precoding
- Resource Elements (RE) mapping
- IQ Compression
- **DM-RS**
  - DM-RS Sequence generation
  - DM-RS Modulation

etc.

Similarly, the AAL_PDSCH_FEC Profile is using the following blocks from that model:

- TB CRC attachment
- CB segmentation + CB CRC attachment
- LDPC encoding
- Rate matching
- CB concatenation

The same specification [12] contains definitions for several O-DU uplink profiles, based on the model "O-DU PHY processing blocks for 5G NR Uplink."

8.5.4 AAL Interface API

8.5.4.1 Introduction

Specification O-RAN.WG6.AAL-GAnP-R003-v08.00 [12] defines two components of the AAL Interface API (Figure 8.17):

- **AALI-C**: A set of common APIs (O-RAN.WG6.AAL-Common-API-R003-v06.00 [11]) for profile-independent aspects, which includes:
  - AALI-C-Mgmt: management interface; administrative operations/actions exposed to IMS
  - AALI-C-App: real-time operations/actions/events exposed to RAN AAL Application
- **AALI-P**: the APIs specific for the AALprofile

Figure 8.17 contains the following entities:

- HW Accelerator, compliant with O-RAN specifications
- **SMO**: O-RAN Service Management and Orchestration framework; contains Network Function Orchestrator (NFO); see Chapter 6
- **IMS**: Infrastructure Management Services, the function targeting deployment and management of the cloud infrastructure (Section 8.3.2)

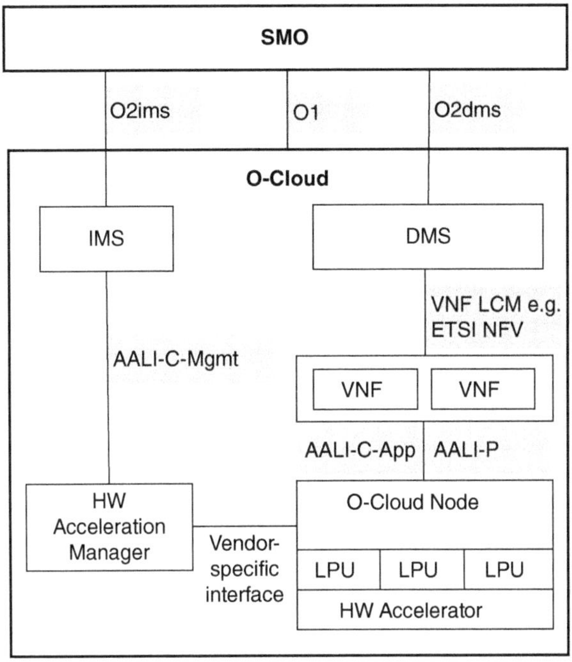

Figure 8.17 AAL Interface components. (*Source:* Vladimir Yanover.)

- **DMS**: Deployment Management Services, the function targeting lifecycle management of virtualized (containerized) deployments on the cloud infrastructure, including LCM of VNFs and NSs (Sections 8.3.3)
- HW Acceleration Manager, the endpoint for the AALI-C-Mgmt interface
- VNFs deployed on the O-Cloud, which are using the HW Accelerator
- O-Cloud Node where the HW Accelerator is installed
- AAL-LPUs representing the resources related to the HW Accelerator

Table 8.5 provides a summary of the interfaces and APIs displayed in Figure 8.17.

8.5.4.2 AALI-C-Mgmt: Common Administrative Operations/Actions/Events

The AALI-C-Mgmt API (O-RAN.WG6.AAL-Common-API-R003-v06.00 [11]) is exposed by the Accelerator Manager and consumed by the Infrastructure Management Service block (Figure 8.17).

The AALI-C-Mgmt supports the following operations:

- **Inventory Management operations**

Table 8.5 Summary of the interfaces and APIs

Interfaces, APIs	Description
AALI-C-Mgmt	The API supporting common administrative operations/actions/events from the Accelerator Manager toward the O-Cloud Infrastructure Management Service. This API is specified in O-RAN.WG6. AAL-Common-API-R003-v06.00 [11]
AALI-C-App	The API supporting real-time operations/actions/events exposed to RAN AAL Application. This API is specified in O-RAN.WG6. AAL-Common-API-R003-v06.00 [11]
AALI-P	AAL profile-specific API; addressed in O-RAN specifications, every of which is targeting certain category of profiles, e.g. High PHY (O-RAN. WG6.AAL-HI_PHY-R003-V05.00 [13]) or FEC (O-RAN.WG6. AAL-FEC.0-v03.00 [22])
O1	FCAPS management of VNFs and cloud components; see O-RAN.WG10. O1-Interface.0-R003-v12.00 [23], O-RAN.WG4.MP.0-R003-V14.00 [15], O-RAN.WG5.O-DU-O1.0-R003-v09.00 [24], O-RAN.WG6. AAL-Common-API-R003-v06.00 [11]
O2ims	The part of the O2 interface, related to Infrastructure Management Services, carrying logical services provided by the O-Cloud, which allow resource management with the purpose of orchestration of the O-Cloud itself
O2dms	The part of the O2 interface, related to Deployment Management Services; see Section 3.7.2 in O-RAN.WG6.O2-GA&P-R003-V06.00 [9]. The O2dms interface enables the SMO to trigger the LCM procedures, such as deploying, terminating, scaling an O-RAN Cloudified NF instance
VNF LCM	Denotes third-party open interface for VNF life cycle management. For example, in the ETSI NFV environment, this interface corresponds to the reference point Ve-Vnfm-vnf between the VNFM and VNF (ETSI GS NFV-MAN 001 [20]), which supports VNF life cycle management actions, such as VNF instantiation, update (e.g. configuration update), scaling out/in and up/down, termination.

The function getAalAccelInfo is used to query information about the HW Accelerator, such as capabilities of the AAL-LPU, AAL-Profiles supported by these AAL-LPU(s), and supported configurations.

- **Configuration Management operations**

  Configure the HW Accelerator and AAL-LPU with setAalAccelConfig

- **Performance Management operations**

  Configure generation and reporting of information associated with AAL-LPU(s), such as performance measurements, counters, KPIs; configure PM collection interval

- Fault and Alarms operations with getAalHwAccelFaults call, which returns the list of currently active faults
- Software/Firmware upgrade operations
- **Operational commands**
  - Query the status of the HW accelerator and the AAL-LPUs
  - Control the state of the AAL-LPU(s): restart, reset
  - Set the power cycle state
  - Start/Stop the AAL-LPU
  - Initiate Test

The following are examples of Inventory Management information for every HW Accelerator:

- Identifier of the accelerator
- List of AAL-LPUs; every element of the list contains for example:
  - Information of the AAL-LPU software image
  - A list of AAL-Profiles supported by the AAL-LPU
  - ...
  - etc.

The following are examples of configuration settings for specific AAL-LPU.

- List of supported AAL-Profiles
- Memory size to be allocated to the AAL-LPU
- Number of compute slices, etc.

AALI-C-Mgmt operations can be triggered by O2 operations. Figure 8.18 shows an example of such triggering with the goal to update the inventory database in the

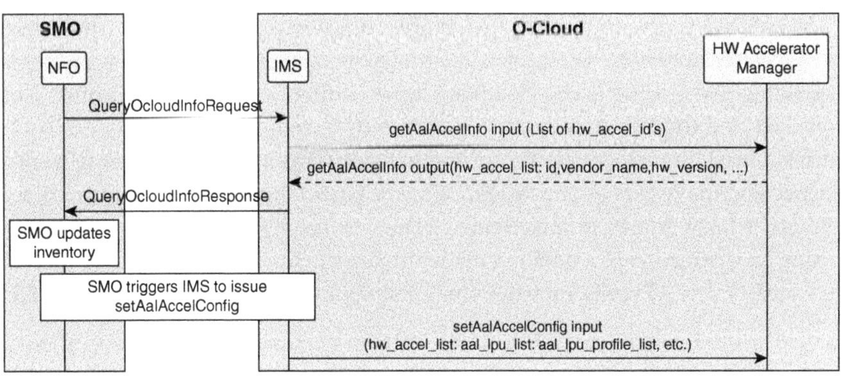

Figure 8.18 AALI-C-Mgmt operations example: inventory update and Accelerator configuration. (*Source:* Vladimir Yanover.)

SMO and configure the HW accelerator. The scenario includes the following actors and roles:

- SMO functions FOCOM/NFO (see Chapter 6)
- IMS, DMS (see Section 8.3)
- HW Accelerator Manager
- HW Accelerator

The example includes the following steps:

- The SMO (NFO) sends over O2 interface QueryOcloudInfoRequest, a query on the HW accelerator, which carries identifiers of HW accelerators
- The IMS queries HW Accelerator Manager for the accelerator information using, the getAalAccelInfo via AALI-C-Mgmt interface. The query contains a list of hw_accel_id identifying accelerators in question
- The IMS receives a response from the HW Accelerator Manager with the requested information: the list of accelerators hw_accel_list with their attributes: accelerator id, the vendor name, HW version, etc.
- The IMS responds to the SMO with QueryOcloudInfoResponse
- SMO receives response from the IMS and updates the inventory using the received information
- Based on the planned use of the accelerators in the deployment, the SMO, with assistance of the IMS, configures the Accelerator via the AALI-C-Mgmt interface using the setAalAccelConfig. For example, AAL-LPU capabilities and Profile images can be configured

8.5.4.3 AALI-C-App: Common Operations/Actions/Events

This section describes operations on the AALI-C-Application interface, which can be performed by the RAN AAL Application.

In addition to operations with AAL Profile Instances and AAL Queues, the AALI-C-App interface enables control of the transport carrying data between the corresponding SW module in the cloudified network function ("AAL Application") on one side and the HW Accelerator together with accompanying software, such as a driver ("AAL implementation"), on another side. This includes the data to be processed and the results of processing. Different types of transport are supported, e.g. Ethernet LAN and PCIe interconnect. The Transport Abstraction Framework is using an abstraction of a buffer; it allows management of buffer pools and individual buffers. The API calls implementing operations with buffers can be of two kinds:

- Synchronous (Blocking) API – the call returns only after the operation is completed
- Asynchronous (Non-Blocking) API – returns immediately. There is a possibility of registering for a callback

The operations on the AALI-C-Application interface can be categorized as follows:

- **Inventory Services**
  Example: getAalLpuInfo to query the capabilities of the AAL-LPU, such as AAL Profile types supported by the AAL-LPU
- **Profile Instance life cycle management**
  For example, operations that create/delete/configure an AAL Profile Instance, based on specific AAL Profile type: createAalProfileInstance, deleteAalProfileInstance, getAalProfileInstanceConfig, setAalProfileInstanceConfig
- **State control operations**
  For example, operations that start/stop AAL Profile instances and AAL Queues: AalProfileInstance, stopAalProfileInstance, startAalQueue, stopAalQueue
- **Buffer operations**
  For example, operations that create/destroy buffer pools: createBufferPool, destroyBufferPool. The operations that allocate a buffer out of the pool and free the buffer: allocBuffer, freeBuffer. The operations of queueing/dequeuing buffers for processing: sendBuffer, receiveBuffer. There are also operations that return the buffer's attributes: getBufferAddr, getBufferSize.

References

1 ETSI GR NFV 003. Network Functions Virtualisation (NFV); Terminology for Main Concepts in NFV.
2 3GPP TS 38.401. NG-RAN; Architecture description.
3 3GPP TR 38.801. Study on new radio access technology: Radio access architecture and interfaces.
4 O-RAN.WG4.CUS.0-R003-v14.00. Control, User and Synchronization Plane Specification
5 O-RAN.WG6.CADS-v06.00. Cloud Architecture and Deployment Scenarios for O-RAN Virtualized RAN.
6 O-RAN.WG6.ORCH-USE-CASES-R003-v09.00. Cloudification and Orchestration Use Cases and Requirements for O-RAN Virtualized RAN.
7 O-RAN.WG6.O2DMS-INTERFACE-ETSI-NFV-PROFILE-R003-v06.00. O2dms Interface Specification: Profile based on ETSI NFV Protocol and Data Models.
8 O-RAN.WG6.O2DMS-INTERFACE-K8S-PROFILE-R003-v04.00
9 O-RAN.WG6.O2-GA&P-R003-v06.00 O2. Interface General Aspects and Principles.
10 O-RAN.WG6.O2IMS-INTERFACE-R003-v05.00. O2ims Interface Specification.
11 O-RAN.WG6.AAL-Common-API-R003-v06.00. Acceleration Abstraction Layer Common API.

12 O-RAN.WG6.AAL-GAnP-R003-v08.00 O-RAN. Acceleration Abstraction Layer General Aspects and Principles.

13 O-RAN.WG6.AAL-HI_PHY-R003-v05.00. Acceleration Abstraction Layer High-PHY Profiles.

14 ETSI GR NFV-IFA 046 V5.1.1. Network Functions Virtualisation (NFV) Release 5; Architectural Framework; Report on NFV support for virtualisation of RAN.

15 O-RAN.WG4.MP.0-R003-v14.00 O-RAN WG4: O-RAN. Fronthaul Management Plane Specification, v10.0

16 eCPRI Specification V2.0 (019-05-10). Common Public Radio Interface: eCPRI Interface Specification.

17 IEEE Std. 1914.3: IEEE Standard for Radio over Ethernet Encapsulations and Mappings.

18 IEEE. Std 1588-2019 Standard for a Precision Clock Synchronization Protocol for Networked Measurement and Control Systems, November 2019.

19 ITU-T Recommendation G8275.1/Y.1369.1. (2020): Amendment 2 (06/2021) Precision Time Protocol Telecom Profile for phase/time synchronization with full timing support from the network.

20 ETSI GS NFV-MAN 001. Network Functions Virtualisation (NFV); Management and Orchestration.

21 ETSI GS NFV-IFA 002 V2.4.1. Network Functions Virtualisation (NFV) Release 2; Acceleration Technologies; VNF Interfaces Specification.

22 O-RAN.WG6.AAL-FEC.0-v03.00. Acceleration Abstraction Layer FEC Profiles.

23 O-RAN.WG10.O1-Interface.0-R003-v12.00 O-RAN. Operations and Maintenance Interface Specification

24 O-RAN.WG5.O-DU-O1.0-R003-v09.00 O1 Interface specification for O-DU

9

Network Automation in O-RAN

CONTENTS

Acronyms and Abbreviations

AI	Artificial Intelligence
C-SON	Centralized SON
CA	Carrier Aggregation
CN	Core Network
CQI	Channel Quality Indicator
CU	RAN Central Unit
D-SON	Distributed SON
DC	Dual Connectivity
DL	Downlink
DRB	Data Radio Bearer
DRX	Discontinuous Reception
DU	RAN Distributed Unit

Open RAN: Technology and Ecosystem, First Edition. Vladimir Yanover.

eNB	An abbreviation of eNodeB; LTE Base Station defined in 3GPP
gNB	An abbreviation of gNodeB; 5G Base Station defined in 3GPP
HO	Handover
H-SON	Hybrid SON
LCM	Life Cycle Management
MIMO	Multiple Inputs Multiple Outputs schema for antenna system
MAC	Medium Access Control
ML	Machine Learning
MnS	Management network Services
MU-MIMO	Multi-User MIMO
Near-RT RIC	Near-Real-Time Radio Intelligent Controller
NF	Network Function
Non-RT	Non-real time
Non-RT RIC	Non-Real-Time Radio Intelligent Controller
NRT	Non-Real Time
NSI	Network Slice Instance
NSSI	Network Slice Subnet Instance
O1	O-RAN management interface (OAM, FCAPS)
O2	O-RAN cloud management interface
O-Cloud	Cloud defined in O-RAN, which exposes O2 interface
O-CU	O-RAN defined Central Unit
O-CU-CP	O-RAN defined Central Unit – Control Plane
O-CU-UP	O-RAN defined Central Unit – User Plane
O-DU	O-RAN defined Distributed Unit
O-eNB	O-RAN conformant LTE Base Station, which, in particular, supports E2 interface
OFH	Open Fronthaul interface
O-gNB	O-RAN conformant 5G Base Station, which, in particular, supports E2 interface
O-RU	O-RAN defined Radio Unit
OAM	Operations, Administration, and Management
PDCP	Packet Data Convergence Protocol
QoS	Quality of Service
RAN	Radio Access Network
RLC	Radio Link Control
RRM	Radio Resources Management
PM	Performance Management
PRB	Physical Resource Block in LTE and 5G
QoE	Quality of Experience
RAN	Radio Access Network
RAT	Radio Access Technology

RLC	Radio Link Control
RT	Real Time
RU	Radio Unit
RRU	Remote Radio Unit
RSRP	Reference Signal Received Power
RSRQ	Reference Signal Received Quality
SDU	Service Data Unit
SLA	Service Level Agreement
SMO	Service Management and Orchestration framework
SON	Self-Organized Network
SST	Slice/Service Type
TN	Transport Network
UE	User Equipment
UL	Uplink
V2X AS	V2X Application Server

9.1 Introduction. SON Solutions Defined in 3GPP

Importance of network automation in the context of profitability is widely recognized; for example, the TM Forum report [1] links network slicing profitability to fully automated end-to-end service operations.

This chapter describes O-RAN technology advances, which leverage network automation/optimization as compared to "classic" optimization solutions, such as centralized SON (C-SON) and Distributed SON (D-SON). Section 9.2 outlines basics of RAN optimization in O-RAN architecture. Sections 9.3, 9.4, 9.5, 9.6, 9.7, and 9.8 identify important automation/optimization scenarios, in some cases together with architectural aspects of AI/ML use.

The 3GPP specification TS 28.313 [2] defined Centralized, Distributed, and Hybrid SON solutions.

Centralized SON is defined as RAN management and optimization solution with optimization algorithms running on top of RAN management services (3GPP TS 28.530 [3], TS 28.531 [4], TS 28.532 [5]). Figure 9.1 shows 3GPP management system, which is a consumer of network Management Services (MnS) provided by the RAN and Core Network Functions. The management system is using Performance Assurance Management Services for collection of network performance data from the NFs (3GPP TS 28.552 [6]). In the case when optimization action is needed, the management system is using Provisioning MnS for changing the configuration of the network to steer its behavior to the desired direction.

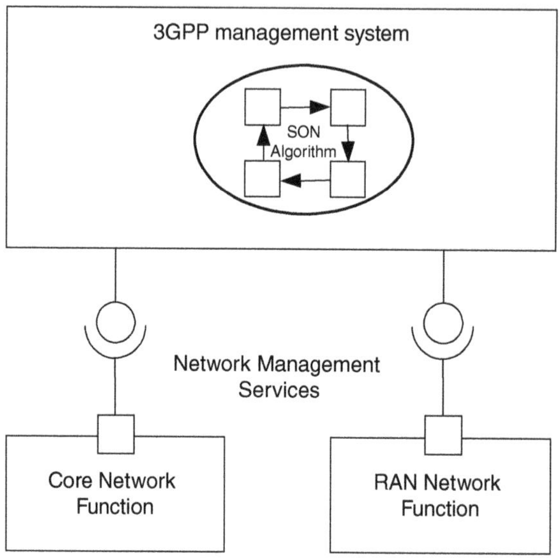

Figure 9.1 Centralized SON (C-SON) architecture. (*Source:* Vladimir Yanover.)

A fundamental C-SON issue is lack of real-time capabilities. The reason is that C-SON data collection and optimization actions go over the management plane; therefore, the response time cannot be low.

In the Distributed SON (D-SON) solution, a separate instance of SON algorithm is running in every Network Function (NF); see Figure 9.2. Every instance collects locally available network performance data, makes optimization decisions, and executes optimization actions. To achieve network optimization at the RAN area level, the instances of the D-SON algorithm communicate to each other via Xn/X2 interfaces. 3GPP standardization for RAN D-SON is mainly focused on Xn/X2 protocols of communication between RAN nodes. 3GPP generally does not specify algorithms, but the Xn information model in 3GPP TS 38.423 [7], allows to outline the family of algorithms, which can be used in the 5G RAN D-SON.

Figure 9.2 shows RAN NFs, hosting D-SON instances together with the interfaces they expose to the management system. These interfaces enable OAM management of D-SON instances, such as switching on/off the D-SON instance, setting D-SON policies, and targets, delivery of enrichment information to the D-SON function.

A fundamental D-SON issue is that D-SON is based on multiple SON algorithm instances, running on multiple separated RAN nodes. It means that for a RAN area where multiple gNodeBs are deployed, the D-SON algorithm essentially is a distributed algorithm. Distributed algorithms can be efficient area wide, but to achieve this, the algorithm instances should simultaneously converge in the whole area.

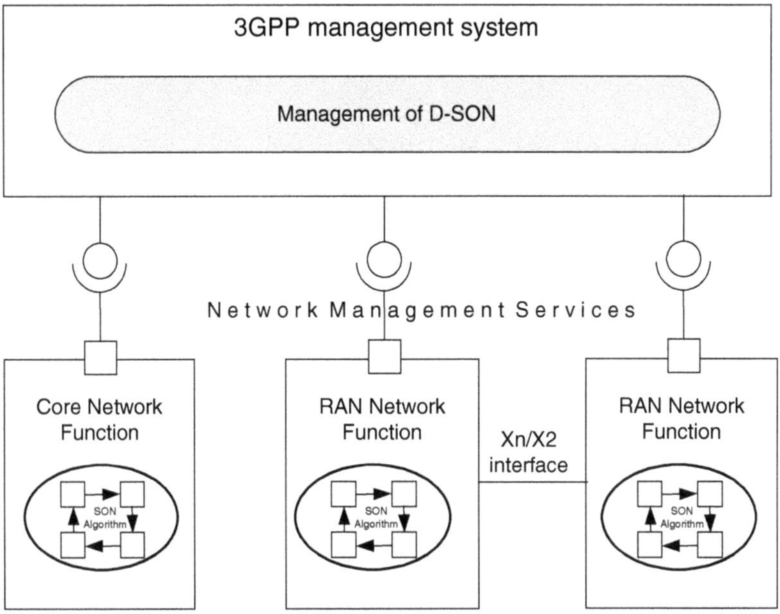

Figure 9.2 Distributed SON (D-SON) architecture. (*Source:* Vladimir Yanover.)

Such simultaneous convergence is difficult to achieve, and it may essentially add to the SON response time.

Figure 9.3 shows Hybrid SON (H-SON) model, where SON algorithms are simultaneously executed at the NF level and above. Network performance data is collected and processed at both levels. In some cases, the 3GPP management system may make decisions and execute optimization actions; in other cases it can modify the configuration parameters of the local SON instances running in the NFs.

3GPP TS 28.312 [8] defines intent-driven management services for mobile networks. The intent-based approach allows to instruct the managed system about the operation goals and the context in which these goals should be achieved. The specification defines intent as "expectations including requirements, goals and constraints given to a 3GPP system, without specifying how to achieve them." The structure of the intent includes Intent Expectation and Context. Example of intent from 3GPP TS 28.312 [8]: ensure that handover failure rate <2% (Intent Expectation) in the cells with load >80% (Intent Context).

In line with the service-based approach, 3GPP defines intent-driven Management Services (MnS). At the step of service setup, the consumer of the MnS informs the service provider of the goals to be achieved. Further communication between the Service Consumer and Service Provider may include performance reporting to

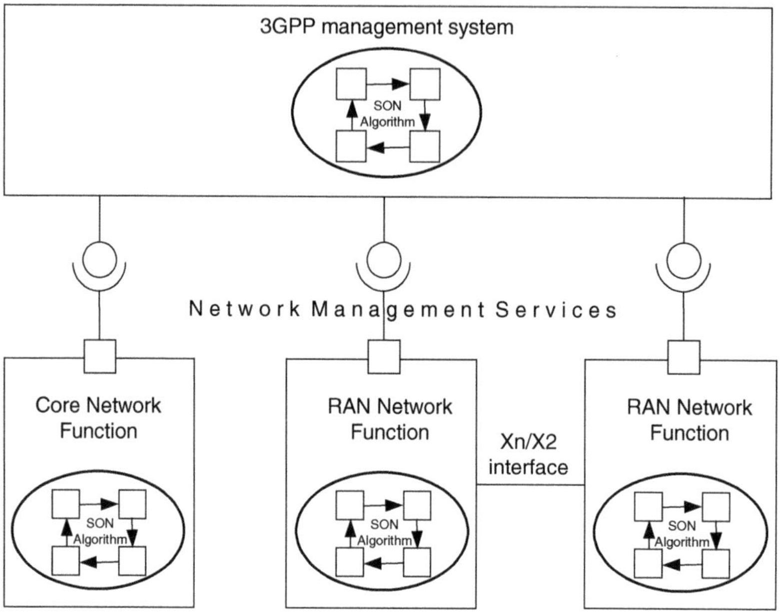

Figure 9.3 Hybrid SON (H-SON) architecture. (*Source:* Vladimir Yanover.)

update the Service Consumer on the service status, i.e. whether the requested goals are achieved.

9.2 Basics of Network Optimization in O-RAN

Specification O-RAN.WG3.UCR-R003-v05.00 [9] outlines O-RAN approach to RAN control and optimization, as applied to several basic use cases. Similarly to 3GPP H-SON, a two-layer approach can be used with Non-Real-Time Radio Intelligent Controller (Non-RT RIC) at the upper layer and Near-Real-Time Radio Intelligent Controller (Near-RT RIC) at the lower layer.

The O-RAN two-layer approach in general combines advantages of C-SON and D-SON. Non-RT RIC functionality allows implementation of management and optimization capabilities provided by 3GPP C-SON. Non-RT RIC can be used as a centralized C-SON platform with network data collection and optimization actions going via the O1 interface. The Near-RT RIC enables near-real-time control loop with the response as fast as in the D-SON. But instead of distributed multiple D-SON instances, the Near-RT RIC-based solution has a single instance of optimization algorithm running on the Near-RT RIC. This eliminates the issue of convergence of distributed SON algorithm.

In addition to combined C-SON and D-SON operational advantages, O-RAN architecture provides additional level of flexibility enabling non-real-time and real-time RAN programming, which can be done by the network operator after O-RAN equipment has already been deployed:

- Possibility of setting intelligent non-real-time policies from the Non-RT RIC to the Near-RT RIC via the A1 interface; support of third-party applications (rApps) running on the Non-RT RIC platform
- Possibility of setting intelligent real-time policies from the Near-RT RIC to E2 Nodes via the E2 interface; support of third-party applications (xApps) running on the Near-RT RIC platform.

It can be added that the O-RAN architecture enables data collection, control and optimization with resolution of single UE and even single QoS Flow.

In the O-RAN architecture, Service Management and Orchestration framework (SMO)/Non-RT RIC can be made a provider of intent-based Management Service (MnS). If such MnS is exposed, the network operator will be able to provide a guidance to the Non-RT RIC on desired RAN performance and/or desired optimization criteria. This guidance can influence the A1 policies, generated by the Non-RT RIC for subordinated Near-RT RIC(s).

Figure 9.4 shows O-RAN optimization architecture, which can be applied to various use cases. The O-RAN architecture allows non-real-time control loop, also called "slow loop" in O-RAN specifications, with the time constraint 1 second or more. The architecture also supports near-real-time control loop, also called "fast loop" in O-RAN specifications, with time constraint between 10 ms and 1 second. The E2 POLICY Service allows real-time control with time constraint less than 10 ms.

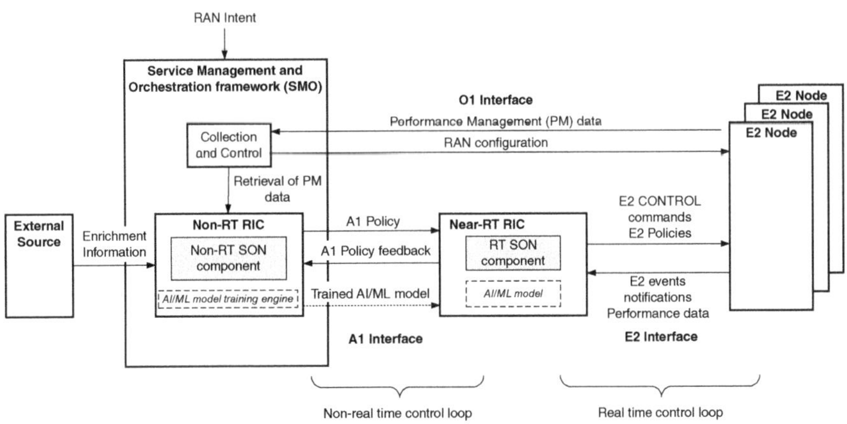

Figure 9.4 O-RAN optimization architecture. (*Source:* Vladimir Yanover.)

The E2 Node can be complete O-gNB or one of RAN components defined in O-RAN: O-CU, O-CU-CP, O-CU-UP, and O-DU. A Near-RT RIC can control multiple E2 Nodes of same kind or of different kinds. Optimization goals (intents) can be set in the Non-RT RIC via the management interface. SMO is collecting network performance data from RAN components, using FCAPS interface O1 [10]. The O1 procedures and information models essentially follow 3GPP management specifications, such as 3GPP TS 28.530 [3], TS 28.531 [4], TS 28.532 [5], TS 28.552 [6]. Collected performance data can be retrieved by the Non-RT RIC as an input to its non-real-time optimization function. Enrichment Information is another input, coming from external sources, such as Application Servers. After analysis of the received data, Non-RT RIC may decide that an optimization action is needed. The action may include modification of the OAM parameters in the involved network functions and/or creation or modification of the non-real-time policy ("A1 Policy") to be enforced by the Near-RT RIC. The Non-RT RIC is using the A1 interface for sending the policy information to the Near-RT RIC (see Chapter 6 and [11]). To notify of the A1 policy status change (e.g. ENFORCED to NOT_ENFORCED), the Near-RT RIC may deliver A1 Policy Feedback to the Non-RT RIC.

For execution of the received A1 Policy, the Near-RT RIC configures subscriptions for receiving from E2 Nodes real-time network performance data and notifications of the E2 events (Chapter 7). The performance data and notifications are delivered via the E2 interface. The Near-RT RIC determines required optimization actions and sends the corresponding real-time instructions to the E2 Nodes using E2 CONTROL Service commands (Chapter 7). Another possibility is that the Near-RT RIC sets E2 Policies at the E2 Nodes, using the E2 POLICY Service. The E2 policy actions are executed by the E2 Node itself, when triggered by local events or conditions.

The sequence diagram in Figure 9.5 shows the flow of the O-RAN optimization process where both Non-RT RIC and Near-RT RIC are involved.

The process begins when OAM configures optimization targets (intents) in the Non-RT RIC. The SMO Collection and Control entity subscribes to the RAN components (E2 Nodes) for delivery of Performance Management (PM) data. The OAM functions in the SMO configure basic parameters in E2 Node(s) through the O1 interface. Subscription and delivery of the collected data go over the O1 management interface. The Non-RT RIC retrieves the collected data from the Collection and Control entity. In addition, Non-RT RIC may receive and analyze Enrichment Information (EI), such as application level information on utilization of specific QoS Flows. The Non-RT RIC can use collected network data and Enhancement Information as an input for training of the AI/ML models. The trained models subsequently will be used in the Non-RT RIC and/or Near-RT RIC for network optimization.

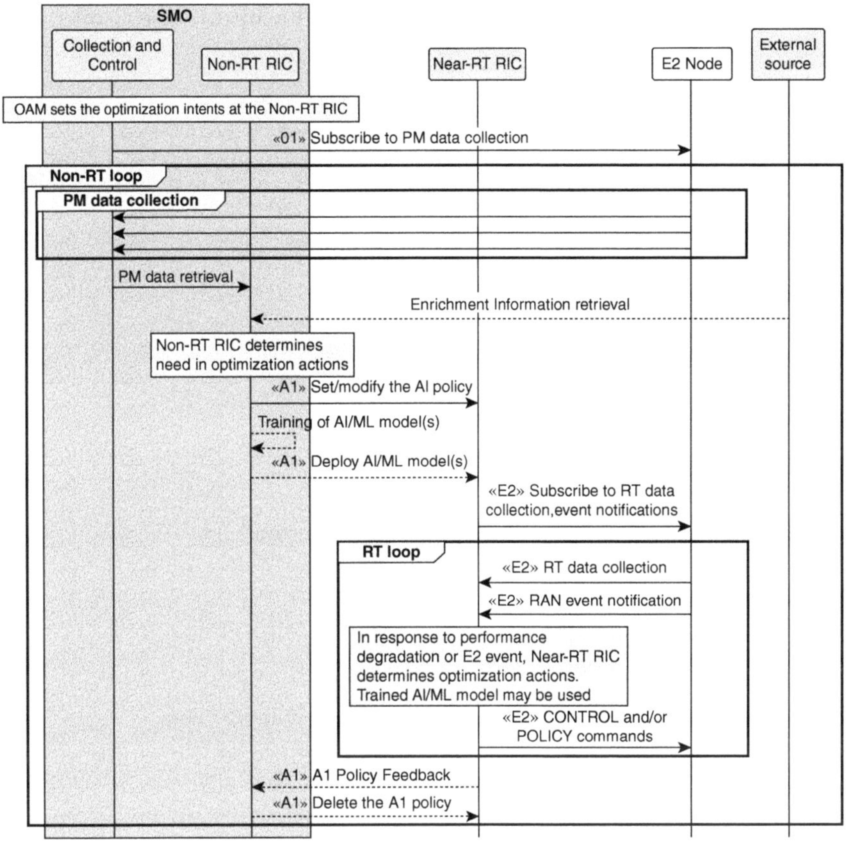

Figure 9.5 O-RAN optimization flow. (*Source:* Vladimir Yanover.)

Based on the collected network data, Non-RT RIC may determine the need in optimization actions. Then the Non-RT RIC sets or modifies the corresponding A1 Policy at the Near-RT RIC via the A1 interface. If AI/ML inference engine is in use in the Near-RT RIC, the Non-RT RIC may also update the trained model at the Near-RT RIC. For execution of the A1 policy, the Near-RT RIC subscribes to relevant E2 Nodes for real-time data collection (E2 REPORT Service) and/or real-time RAN event notifications (E2 INSERT Service). The E2 Nodes report this information via REPORT and INSERT services periodically or event-triggered (Chapter 7, [12, 13]).

In case the received network data or specific RAN events indicate deviation from the RAN performance targets, the Near-RT RIC may decide on optimization actions and execute them using E2 CONTROL or E2 POLICY commands sent via the E2 interface. The decision can be based on the outcome of AI/ML inference function running at the Near-RT RIC.

O-RAN specifications [9, 14, 15] define optimization use cases, such as:

- Context-Based Dynamic HO Management for V2X
- Flight-Path-Based Dynamic UAV Radio Resource Allocation
- Radio Resource Allocation for UAV Application Scenario
- QoE Optimization
- Traffic Steering
- Massive MIMO Beamforming Optimization
- RAN Sharing
- QoS-Based Resource Optimization
- RAN Slice SLA Assurance
- Multi-vendor Slices
- Dynamic Spectrum Sharing
- NSSI Resource Allocation Optimization
- Local Indoor Positioning in RAN
- O-RAN Signaling Storm Protection
- Integrated SON Function within the O-RAN framework
- Shared O-RU
- Energy Saving
- MU-MIMO Optimization
- Industrial vision SLA Assurance

The following sections describe several use cases in more detail.

9.3 Traffic Steering Use Case

This section outlines a solution for the Traffic Steering use case in O-RAN architecture.

The term "Traffic Steering" refers to a category of network optimization actions, which include steering of the User Plane traffic from one path to another for improvement of overall network resources utilization and/or improvement of service quality. One example is traffic steering between frequency channels, with or without carrier aggregation. In this case the role of network resources is played by frequency channels. Another example is traffic steering between Primary Cell and Secondary Cell(s) in Dual Connectivity; in this case the resources (PRBs) of the corresponding cells are in play.

The O-RAN specification O-RAN.WG3.UCR-R003-v05.00 [9] describes several types of Traffic Steering solutions:

1) The "Baseline" solution, which follows the C-SON approach with optimization algorithm running in the SMO. Network data collection and optimization

actions go over the O1 interface. In this option, only non-real-time control loop is used; the Near-RT RIC is not involved

2) The "Background" solution using both non-real-time and real-time control loops. The Non-RT RIC hosts the non-real-time component of the optimization algorithm; it is using O1 interface for configuration of the Near-RT RIC to set it up for desired behavior. The Near-RT RIC, in turn, activates E2 REPORT Service to monitor the E2 Nodes performance indicators. The E2 INSERT and E2 CONTROL Service mechanisms can be used to keep the E2 Nodes in line with the desired behavior; alternatively, E2 POLICY Service can be used for that purpose. The process flow follows Figure 9.5, with correction that instead of A1 operations shown in the figure, the Non-RT RIC applies O1 operations.
3) The "A1-Policy-based" solution, where the Non-RT RIC uses A1 interface for setting an A1 Policy at the Near-RT RIC. The A1 Policy directs the Near-RT RIC behavior according to the optimization targets. The Near-RT RIC uses E2 REPORT service and E2 INSERT service to monitor the E2 Nodes. E2 CONTROL and E2 POLICY service mechanisms are used to execute the optimization actions toward the E2 Nodes, according to the A1 Policy. The process flow is shown in Figure 9.5.

The following are some categories of network data, which can be used by the Non-RT RIC and Near-RT RIC in Traffic Steering:

- Performance measurements ("counters") and KPIs defined in 3GPP management specifications. This includes such performance data as DL/UL PRB usage statistics per use category (e.g. PDCCH; see 3GPP TS 38.300), PDCP SDU data volume, and handover success ratio. Such performance indicators usually are aggregated over RAN cells, network slices, QoS categories, etc.
- Cell load statistics, usually identical to those delivered over X2/Xn interface, for example, PRB utilization, number of active UEs, bearer metrics, etc.
- Real-time information of events, such as transmission and reception of protocol messages at the interfaces NG, Xn, F1, and RRC, including the content of the messages. Example of the content: RRC measurements reports information delivered to RAN from the UEs, such as RSRP/RSRQ/CQI measurement results for serving and neighboring cells with or without CA and DC, both intra-RAT and inter-RAT
- Per-UE performance indicators, such as PDCP/RLC/MAC level throughput and latency
- UE Context Information from E2 Nodes, such as
  - UE ID
  - Slice (RAN NSSI) level information
  - DRB level information, such as DRB ID, mapping QoS Flows to DRBs
  - QoS-related information, such as QoS Flow level QoS Parameters
  - UE capabilities, for example CA and DC capabilities

The RAN components (E2 Nodes) may deliver such information via O1 interface to the SMO, where it becomes available to the Non-RT RIC. This information can be also collected by the Near-RT RIC using REPORT services. The RAN events notifications are delivered to the Near-RT RIC via E2 INSERT service (Chapter 7).

The following are examples of E2 Node actions, which can be executed for traffic steering:

- Triggering UE handover from an overloaded cell to a less loaded cell
- Modification of the handover restriction list
- Modification of the IDLE state mobility parameters
- Reconfiguration/enabling/disabling of Carrier Aggregation for specific UEs or a group of UEs
- Reconfiguration/enabling/disabling of Dual Connectivity for specific UEs or a group of UEs

9.4 QoS-Based Resource Optimization Use Case

The O-RAN specification O-RAN.WG3.UCR-R003-v05.00 [9] defines QoS-Based Resource Optimization use case as scenario focused on minimizing usage of resources, while meeting the requirements of reliability, latency, and throughput. It is assumed that the radio resources are allocated to multiple co-existing services, which may have different priorities.

The architecture diagram in Figure 9.4 is fully applicable, as well as the process flow shown in Figure 9.5.

In this use case, Non-RT RIC and Near-RT RIC can use same categories of network data, as in the Traffic Steering (Section 9.3).

The following are examples of E2 Node actions that can be executed:

- **DRB level control actions, based on modification of the DRB QoS properties**:
  - DRB level QoS properties
  - Parameters that control QoS Flows mapping to DRBs
  - Logical channel parameters, such as priority (TS 38.321 [16])
  - RAN admission control parameters
  - Dual Connectivity (DC) DRB parameters, such as the points of the bearer termination
  - Configuration of the legs in the Dual Connectivity (DC), Carrier Aggregation (CA), and DC combined with CA
- **Radio resource allocation control actions, based on modification of such parameters as:**
  - Configuration of DRX, semi-persistent scheduling (SPS), scheduling and rate selection in MAC, Long and Short DRX cycles duration

  - Scheduling Request periodicity configuration
  - Configuration of slice level PRB quota
  - Configuration of CQI table with mapping to target block error rate
- Radio access control actions, based on modification of the RACH Backoff configuration and parameters that control RRC connection establishment, release, and rejection (3GPP TS 38.331 [17])
- **Mobility Control actions, such as:**
  - Triggering UE handover to specifically selected target cell
  - (Re)configuration of handover restrictions
  - (Re)configuration of IDLE state mobility parameters
  - (Re)configuration of Carrier Aggregation state or Dual Connectivity state

9.5 RAN Slice SLA Assurance Use Case

This section describes the RAN Slice SLA Assurance scenario.

Slicing feature was defined by 3GPP as a part of 5G, starting from Release 15; see 3GPP TS 22.261 [18]. With slicing, same network infrastructure can simultaneously support multiple logical networks ("network slices"), which may deliver different services to the customers. Slicing feature supports dynamic creation and configuration of network slices, which helps meeting requirements for future applications, services, and business verticals.

A network slice (more precisely, "Network Slice Instance" or NSI) goes across the network domains: Core Network (CN), the Transport Network (TN), and the Radio Access Network (RAN). The corresponding partitions of the slice are called Network Slice Subnet Instances (NSSIs), i.e. it's CN NSSI, TN NSSI, and RAN NSSI.

According to 3GPP TS 28.530 [3], the key attribute of the NSI is the Slice/Service Type (SST). 3GPP TS 23.501 [19] defines the following SST values:

- **eMBB**: 5G enhanced Mobile Broadband
- **URLLC**: Ultra-Reliable Low Latency Communications
- **mIoT**: Massive IoT
- **V2X**: V2X (Vehicle-to-anything) services
- **HMTC**: High-Performance Machine-Type Communications

Depending on the SST, slice configuration may include different requirements taken from the basic set:

- User density requirement
- Area traffic capacity requirement
- Coverage area requirement
- End-to-end latency requirement

- Mobility requirement
- Service availability requirement
- Service reliability requirement

This set of NSI configuration parameters, with specific numerical values included, sometimes is called "Service Level Agreement" (SLA).

The GSMA document NG.116 [20] defines the structure of Generic Slice Template (GST), which identifies the attributes applicable to certain category of network slices. Along with the Slice/Service Type (SST), there can be attributes that describe supported QoS or mobility type. Filled with specific values, a GST becomes the NEtwork Slice Type (NEST) document. GSMA defined several NESTs targeting different service/slice types:

- Enhanced Mobile Broadband with IMS support
- Ultra-reliable and ultra-low latency communication
- Massive IoT
- High-Performance Machine-Type Communications
- Public Safety
- Enhanced Mobile Broadband with IMS and MPS support
- Vehicle-to-Everything (often referred to as V2X)
- XR (Virtual Reality, Augmented reality, . . .)

O-RAN architecture has definite advantages in control and optimization of network slices in the part of RAN NSSI. The reason is that network slices generally should be supported in a wide area, where a significant number of gNodeBs are deployed. Powerful O-RAN centralized controllers, Non-RT RIC and Near-RT RIC, enable network monitoring in the wide area. For monitoring, O-RAN specifications specify a process for collection of network data in wide area, both in non-real time and real time. In case network slice requirements are not fulfilled, the controllers can execute proper correction actions toward all involved gNodeBs in non-real time or in real time.

In the RAN Slice SLA Assurance use case, Non-RT RIC and Near-RT RIC can use same categories of network data, as in the Traffic Steering (Section 9.3). An especially important role is played by performance indicators aggregated over specific slice(s) identified by the corresponding S-NSSAIs. For example, UE throughput indicator can be aggregated over specific network slice; then all UEs connected to the slice will contribute to that indicator.

Possible Near-RT RIC optimization actions toward E2 Nodes and subsequent actions of the E2 Nodes are same as for Traffic Steering (Section 9.3) and QoS-Based Resource Optimization (Section 9.4), combined.

9.6 Context-based Dynamic Handover Management for V2X Use Case

This section contains description of the use case targeting CONNECTED state mobility for V2X scenario. "V2X" means communication of the type "vehicle-to-anything" (another vehicle, network, server, etc.).

V2X communication in general requires high reliability. One example is delivery of safety-related Cooperative Awareness Messages (CAMs) with information about vehicle state, position, and velocity. The problem addressed in the use case is about communication failures caused by non-optimal handovers, which may significantly affect the network reliability. A UE integrated with the vehicle may travel at high speed, which may result in too frequent handovers, ping-pong effect, handovers to non-optimal (e.g. too distant) cells and other problems. The solution outlined in [15] includes the following:

- V2X Application Server (V2X AS) collects handover performance statistical information, which includes timing and location of the handover and the corresponding radio measurements, e.g. RSRP, RSRQ
- The SMO makes the statistical information available to the Non-RT RIC. This can be done using mechanisms for transfer of Enrichment Information (EI)
- The received information is used as a training set for the handover optimization AI/ML model; training can be done in the Non-RT RIC. The output of the model is a list of optimal handover targets for UEs, located in certain place and reporting the radio measurements within certain range. The trained model is delivered to the Near-RT RIC
- The Near-RT RIC subscribes for reporting of real-time handover information, which includes the location of the UE, and the radio measurements reported by the UE; the E2 INSERT Service can be used for this purpose
- The Near-RT RIC is using the real-time handover information as input to the model for generation of optimal handover targets for every individual UE. This information is communicated in real time to the corresponding O-gNB (O-gNB-CU), using E2 CONTROL service
- The O-gNB (O-gNB-CU) executes the E2 CONTROL command received from the Near-RT RIC

The solution follows the general scheme described in Section 9.2, with Figure 9.4 showing the solution architecture and Figure 9.5 showing the process flow; in both diagrams, V2X Application Server plays the role of External Source.

9.7 QoE Optimization Use Case

This section describes architectural aspects of AI/ML-based control and optimization, on the example of QoE Optimization solution, in which training of the AI/ML model is performed in the Non-RT RIC, while use of the trained model for inference is performed in the Near-RT RIC.

The term Quality of Experience (QoE) refers to estimation of the human user experience, derived from measurement of network/application performance indicators. For example, in the case of video streaming, the perceptual quality of video can be estimated based on video encoding bit rate, packet loss, average throughput, rebuffering frequency, stalling time, and similar indicators. Calculation of QoE can be based on use of empirically proven formulas or AI/ML models specifically trained for this purpose.

The QoE Optimization use case targets demanding 5G applications. One example of it: Virtual Reality with the server running in the network (Edge Cloud). Support of such application requires high bandwidth and low latency. In this case, standard 5G semi-static QoS management may be insufficient. In case of radio link degradation, the Quality of Experience (QoE) for the application will degrade too. A possible solution could be to use the QoE estimation and/or prediction. This allows detection of QoE degradation, which can trigger intervention into regular RRM process, for example, allocation of additional resources to the radio links carrying the VR data.

In the solution, based on the O-RAN architecture, the Near-RT RIC is used for data collection and optimization actions via the E2 interface. The received data (cell level, UE level) is processed using AI/ML models for detection of the traffic type and estimation/prediction of the QoE.

O-RAN architecture support for AI/ML model training and distribution is described in O-RAN.WG2.Use-Case-Requirements-R003-v09.00 [15], Section 4.2.3.1 "Model training and distribution."

Figure 9.6 shows the process flow for the model training, deployment, and use.

The process begins when the SMO Collection and Control function subscribes for QoE-related measurement metrics from the RAN components. After subscription is activated, the SMO starts receiving network measurement reports. It is assumed that the SMO receives QoE related data from the application as well. After sufficient amount of data is collected in the SMO, the Non-RT RIC retrieves the data to use it as a training set for the AI/ML models. The trained AI/ML model(s) are delivered to the Near-RT RIC for use in inference. The O-RAN document O-RAN.WG2.AIML-v01.03 [21] indicates that A1-ML interface can be used for this purpose. After the AI/ML models in the Near-RT RIC are activated, they can be used by optimization modules (xApps) for real-time processing of network performance metrics and application performance metrics. The outcome of the

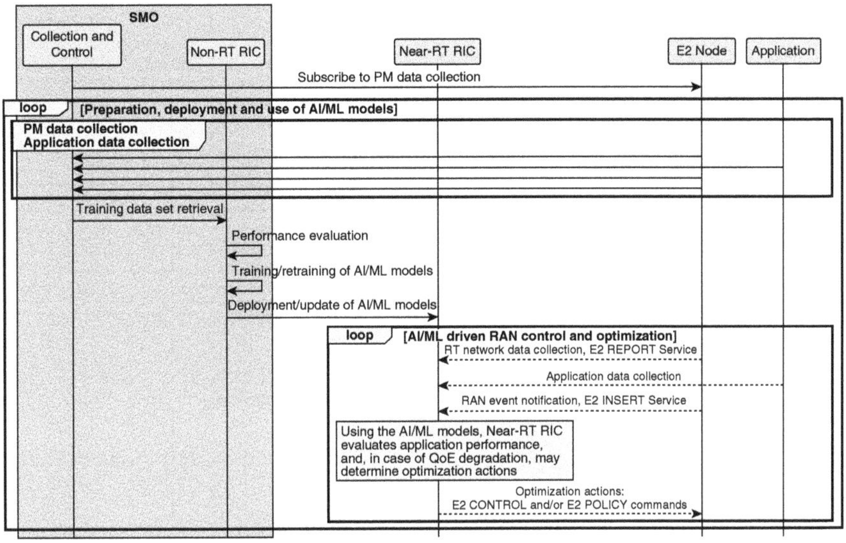

Figure 9.6 QoE optimization. (*Source:* Vladimir Yanover.)

model contains the QoE evaluation and/or prediction. In case evaluated or predicted QoE degrades, the Near-RT RIC may determine required optimization actions, which are executed via the E2 CONTROL or E2 POLICY Services.

9.8 NSSI Resource Optimization Use Case

This section contains description of the NSSI Resource Optimization solution. This solution, described in O-RAN.WG2.Use-Case-Requirements-R003-v09.00 [15], contains two essential elements:

- Optimization of use of RRM resources in the coverage area
- Optimization of use of O-Cloud resources

The first element is not O-RAN specific; it can be implemented in a standard OAM level C-SON. As different from that, the second element is specific for O-RAN architecture, where cloud management is integrated into the system.

Figure 9.7 contains the sequence diagram for NSSI resource optimization. The following entities are involved:

- SMO Collection and Control function
- Non-RT RIC rApp implementing AI/ML-based optimization algorithm

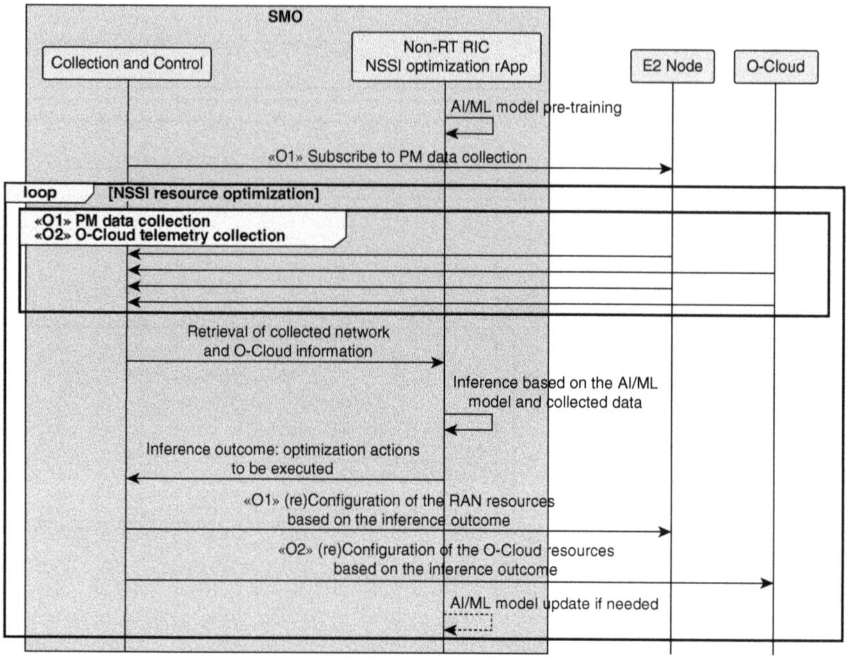

Figure 9.7 NSSI resource optimization. (*Source:* Vladimir Yanover.)

- E2 Node: any of O-CU, O-CU-CP/UP, O-DU. There may be multiple E2 Nodes subordinated to same Non-RT RIC
- O-Cloud instance hosting the RAN VNFs

The process begins with AI/ML model pre-training and subscription to collection of RAN RRM information from relevant E2 Nodes. The NSSI resource optimization loop begins with collection of this data. O-Cloud resource usage information is collected by the SMO via the O2 interface. The data collected in the SMO, can be retrieved by the NSSI optimization application rApp and used as an input to the AI/ML inference process. The rApp feeds the inference results back to the SMO. Based on that, the SMO issues OAM (O1) level instructions to the E2 Nodes and O2 level instructions to the O-Cloud. The O1 instructions provide guidance on (re)configuration of RAN RRM resources use. The O2 instructions provide guidance on (re)configuration of cloud resources. The AI/ML model can be updated (retrained) when needed.

References

1 Ramsay, D. (2024). Closing the loop: CSPs aim to automate service orchestration and assurance.
2 3GPP TS 28.313. Management and orchestration; Self-Organizing Networks (SON) for 5G networks.
3 3GPP TS 28.530. Management and orchestration; Concepts, use cases and requirements.
4 3GPP TS 28.531. Management and orchestration; Provisioning.
5 3GPP TS 28.532. Management and orchestration; Generic management services.
6 3GPP TS 28.552. Management and orchestration; 5G performance measurements.
7 3GPP TS 38.423. NG-RAN; Xn Application Protocol (XnAP).
8 3GPP TS 28.312. Intent driven management services for mobile networks.
9 O-RAN.WG3.UCR-R003-v05.00. Use Cases and Requirements.
10 O-RAN.WG10.O1-Interface.0-R003-v12.00. O-RAN Operations and Maintenance Interface Specification.
11 O-RAN.WG2.A1GAP-R003-v03.02 A1 interface: General Aspects and Principles.
12 O-RAN.WG3.E2AP-R003-v05.00 O-RAN E2. Application Protocol (E2AP).
13 O-RAN.WG3.E2GAP-R003-v05.00. E2 General Aspects and Principles (E2GAP).
14 O-RAN.WG1. Use-Cases-Detailed-Specification-R003-v13.00. Use Cases Detailed Specification.
15 O-RAN.WG2. Use-Case-Requirements-R003-v09.00 Use Cases and Requirements.
16 3GPP TS 38.321. Medium Access Control (MAC) protocol specification.
17 3GPP TS 38.331. Radio Resource Control (RRC) protocol specification.
18 3GPP TS 22.261. Service requirements for the 5G System; Stage 1.
19 3GPP TS 23.501. System architecture for the 5G System (5GS); Stage 2.
20 GSMA NG.116. Generic Network Slice Template Version 9.0 27. April 2023.
21 O-RAN.WG2.AIML-v01.03 AI/ML workflow description and requirements. Technical Report.

10

O-RAN Management

CONTENTS

Acronyms and Abbreviations

A1-P A1 Policy service
A1-EI A1 Enhancement Information service
DMS Deployment Management Services

Open RAN: Technology and Ecosystem, First Edition. Vladimir Yanover.

E2AP	E2 Application Protocol
E2SM	E2 Service Model
EI	A1 Enrichment Information
FCAPS	Fault, Configuration, Accounting, Performance, Security management functions
gNB-CU	gNB – Central Unit
gNB-CU-CP	gNB – Central Unit – Control Plane
gNB-CU-UP	gNB – Central Unit – User Plane
gNB-DU	gNB – Distributed Unit
HTTP	HyperText Transfer Protocol
IE	Information Element
IMS	Infrastructure Management Services
JSON	Java Script Object Notation
KPM	Key Performance Measurements
MnS	Management Service
Near-RT RIC	Near-Real-Time Radio Intelligent Controller
NF	Network Function
NFO	Network Function Orchestration
Non-RT RIC	Non-Real-Time Radio Intelligent Controller
OAM	Operations, Administration and Maintenance
O-Cloud	A cloud platform, supporting hosting of O-RAN network functions
O-CU	O-RAN Central Unit
O-CU-CP	O-RAN Central Unit – Control Plane
O-CU-UP	O-RAN Central Unit – User Plane
O-DU	O-RAN Distributed Unit
O-RU	O-RAN Radio Unit
PNF	Physical Network Function
rApp	Third-party application hosted by the Non-RT RIC
REST	REpresentational State Transfer protocol
TLS	Transport Layer Security protocol
RPC	Remote Procedure Call
VNF	Virtualized Network Function
xApp	Third-party application hosted by the Near-RT RIC

10.1 Introduction

This chapter contains overview of O-RAN management, including O1 interface and Open FH management plane ("M-Plane") interface.

The O1 OAM interface implements Fault, Configuration, Accounting, Performance, Security (FCAPS) management, File Management and Software

Management for O-RAN Network Functions. For more details of the management services supported by O1, see Section 10.2.

Open FH M-plane interface (O-RAN.WG4.MP.0-R003-v14.00 [1]) can be used for O-RU management, either by O-DU or by the SMO. This interface includes implementation of FCAPS functions, File management, and Software management functions for O-RU. It also specifies the carrier management aspects. See Section 10.2.3.1 for more details.

10.2 O1 Interface

10.2.1 Introduction

This section describes basic concepts of the O1 interface responsible for management of the O-RAN nodes: Near-Real-Time Radio Intelligent Controller (Near-RT RIC), O-CU (O-CU-CP, O-CU-UP), O-DU, together with combinations of those (O-RAN. WG10.O1-Interface.0-v12.00 [2], O-RAN.WG10.OAM-Architecture-v11.00 [3]).

In line with 3GPP definitions for 5G, management operations are defined in terms of management services (MnS) defined in 3GPP TS 28.530 [4] and the corresponding information models 3GPP TS 28.541 [5]. O-RAN information models are based on classes and attributes defined in 3GPP TS 28.622 [6], such as ManagedElement and ManagedFunction; see O-RAN.WG10.Information Model and Data Models.0-R003-v07.00 [7].

O-RAN Alliance produced generic management specifications applicable to O-RAN nodes, interfaces, and objects; see Sections 10.2.3, 10.2.4, and 10.2.5. This includes description of procedures, operations, and notifications and information models definitions, which are common for all O-RAN-managed nodes and objects.

In addition to generic specifications, O-RAN Alliance created information models for specific managed nodes and objects. In particular, the following specifications were developed:

- For O-CU-CP, O-CU-UP: O-RAN.WG5.O-CU-O1.0-R003-v07.00 [8]
- For O-DU: O-RAN.WG5.O-DU-O1.0-R003-v09.00 [9]
- For OpenFrontHaul interface: O-RAN.WG4.MP.0-R003-v14.00 [1]

10.2.2 OAM Architecture

Figure 10.1 shows overall O-RAN Logical Architecture, as defined in O-RAN. WG1.OAD-R003-v11.00 [10]; see also Chapter 5. The architecture identifies the following management interfaces:

- O1 interface is used by the O-RAN Service Management and Orchestration Framework (SMO) for OAM management of O-RAN network components: Near-RT RIC, O-CU-CP, O-CU-UP, and O-DU. The O1 Interface supports Fault,

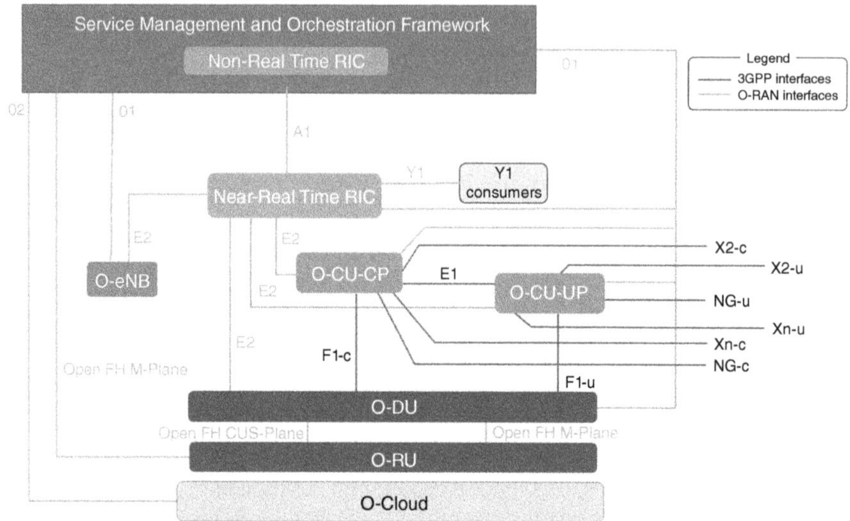

Figure 10.1 Logical Architecture of O-RAN. (*Source:* Reproduced with permission of O-RAN ALLIANCE e.V.)

Configuration, Accounting, Performance, Security (FCAPS) management, file management, and software management. Details of the management services supported by O1 can be found in O-RAN.WG10.O1-Interface.0-R003-v12.00 [2]. The information models for specific O-RAN Network Functions have been developed by different Work Groups; for example, O-RAN.WG5.O-DU-O1.0-R003-v09.00 [9] specifies the information model for the O-DU.

- The Open FH M-plane interface is used for O-RU management by the SMO and/or by the O-DU; it is specified in O-RAN.WG4.MP.0-R003-v14.00 [1]. This specification describes the O-RU management model: hierarchical vs. hybrid.
- O2 interface for management of the O-Cloud. The functions of the O2 interface include management of O-Cloud infrastructure resources and Life Cycle Management (LCM) of the Network Functions hosted by the O-Cloud. See Chapter 8 for more information.

Figure 10.1 shows separate O1 management interfaces to different O-RAN nodes, such as O-CU-CP and O-DU. However, the standard allows combining of Managed Functions into Managed Elements, in which case the Management Element becomes the termination point for the O1 in this specific deployment.

The document “O-RAN Operations and Maintenance Architecture” (O-RAN.WG10.OAM-Architecture-v11.00 [3]) identifies the following building blocks for the management architecture:

- Management Services
  Management Services (MnS) enable management and orchestration of O-RAN Network Functions (NFs); the NFs play the role of MnS Providers. Management

functions, such as SMO, play the role of MnS Consumers. More details on the MnSs can be found in Section 10.2.3

- Managed Elements
  ManagedElement (ME) is the basic Information Object Class (IOC) used in the O-RAN Information Model. It is defined in 3GPP TS 28.622 [6], clause 4.3.3. A ManagedElement represents a ManagedFunction (MF) or a combination of several MFs.
- Managed Functions (MFs) and Network Functions
  3GPP TS 28.622 defines the ManagedFunction IOC in clause 4.3.4.
  The O-RAN MFs include Near-RT RIC, O-CU-CP, O-CU-UP, O-DU, O-RU
- Managed Applications (MAs)
  ManagedApplication is an IOC, which represents the management aspects of the xApp. The O1 interface terminates at the Near-RT RIC platform; the Near-RT RIC platform delegates the management of xApps to external Management Functions. The details on the MA can be found in the specification O-RAN. WG10.Information Model and Data Models.0-R003-v07.00 [7].
- SMO Framework and Non-Real-Time Radio Intelligent Controller (Non-RT RIC)
  See Chapter 6
- Control Loop Support
  See Chapter 9

10.2.3 Management Services in the O-RAN Architecture

10.2.3.1 Introduction

The document O-RAN.WG10.O1-Interface.0-R003-v12.00 "O-RAN Operations and Maintenance Interface Specification" [2] defines the management services supported by the O-RAN management architecture. For such a management service (MnS), an O1-compliant Managed Elements play the role of service providers, and the SMO plays the role of the service consumer. The document describes procedures, operations, and notifications, including Stage 2 level description in terms of NETCONF (IETF RFC 6241 [11]) and YANG (IETF RFC 7950 [12]).

10.2.3.2 Provisioning Management Services

Provisioning Management Services allow configuration of managed objects, such as O-RAN network functions. Another supported capability allows the Provisioning MnS Consumer to subscribe at the MnS Provider for reports on its configuration changes. O-RAN specifications use NETCONF protocol for the Provisioning Management Services. The configuration attributes are modelled using YANG language. O-RAN is using the "YANG/NETCONF solution set" (3GPP term), in line with management specifications 3GPP TS 28.531 [13] and 3GPP TS 28.532 [14]. The supported operations are Creation/Deletion/ Modification of Managed Object instances (MOIs) and Read operation for reading their attributes. The notifications are specified in terms of RESTful/HTTP protocol with YANG modelling and VNF Events Stream (VES) support, according to 3GPP TS 28.532 [14] and OPNFV VES specifications [15].

10.2.3.3 Fault Supervision Management Services

Fault supervision management service enables reporting errors and events from the Fault Supervision MnS Provider to the MnS Consumer. The service allows the MnS Provider to send asynchronous Fault notification event to the Consumer in case when the alarm state changes, i.e. when an alarm occurs or is cleared or changes severity. Fault supervision MnS supports notifications in two formats: VES and SDO; the latter is defined in 3GPP TS 28.532 [14].

In addition, the Fault Supervision MnS consumer can perform operations on the MnS Provider, such as pulling the alarm list. The Fault Supervision MnS is defined in 3GPP specifications TS 28.532 [14], 3GPP TS 28.545 [16], 3GPP TS 28.622 [6], 3GPP TS 28.623 [17].

10.2.3.4 Performance Assurance Management Services

Performance Assurance Management Services enable reporting the network performance data from the MnS Provider to MnS Consumer. The are two types of reporting: file-based reporting and streaming data reporting. The services allow the MnS Consumer to identify the performance indicators to be reported and reporting parameters, such as reporting frequency. For related definitions, O-RAN is using 3GPP specifications 3GPP TS 28.532 [14], 3GPP TS 28.622 [6], 3GPP TS 28.623 [17].

The O-RAN specifications allow delivery of all performance indicators defined in 3GPP TS 28.552 [18]. Using same mechanism, it's possible to add performance indicators defined exclusively in O-RAN or even defined by specific vendor.

In case a large volume of performance indicators should be transferred, asynchronous streaming data reporting can be used. The Performance Assurance Service supports delivery of multiple simultaneous streams, every stream produced by the corresponding Measurement Job. The process includes establishment of the WebSocket connection.

10.2.3.5 Trace Management Services

Trace management in O-RAN follows 3GPP specification TS 28.532 [14].

Trace Management Services ("Trace MnSs") allow collection of information encoded in the protocol messages at different network layers at the Trace MnS Provider; the information is delivered to the Trace MnS Consumer (3GPP TS 32.421 [19]). The Trace MnS configuration includes the trace scope: which Network Elements and which signaling interfaces are subject to tracing. Examples of data to be collected: the Minimization of Drive Testing (MDT) data, Radio Link Failure (RLF) reports, RRC Connection Establishment Failure (RCEF) reports. The configuration of the Trace MnS includes the Trace depth parameter that identify the level of details in the trace data. The Trace Management Services Provider transfers the Trace content in the format of files or in the format of streaming trace records. The trace sessions are linked to the corresponding Trace Jobs running on the Trace Management Services Provider.

10.2.3.6 Other Management Services

O-RAN specification O-RAN.WG10.O1-Interface.0-R003-v12.00 [2] defines the following O-RAN management services; where relevant, it refers to 3GPP specifications.

10.2.3.6.1 File Management Service A File Management Service Provider can notify the Service Consumer of new available files, which are of interest for the Consumer. The Consumer may query about the files and request transfer of the files. In the part of File Management Services, O-RAN follows definitions in 3GPP TS 32.342 [20] and 3GPP TS 32.346 [21].

10.2.3.6.2 Heartbeat Management Service A Consumer of the Heartbeat MnS configures the heartbeat services on the Heartbeat MnS Provider, after which the Provider sends heartbeats to the Consumer. O-RAN is using definitions from the following 3GPP specifications: 3GPP TS 28.532 [14], 3GPP TS 28.622 [6], 3GPP TS 28.623 [17].

10.2.3.6.3 PNF Startup and Registration Management Service These services are defined in the 3GPP specifications 3GPP TS 28.314 [22], 3GPP TS 28.315 [23], and 3GPP TS 28.316 [24].

10.2.3.6.4 PNF Software Management Service O-RAN Software Management Services aim support of software download and installation, followed by validation and activation of the software package. The attributes of the software package include the name, version, and file list.

10.2.3.6.5 PNF Reset Management Service The PNF Reset Management Service allows the MnS Consumer to reset selected HW components at the MnS Provider. Two types of the reset command are defined: conditional and forced. The MnS Provider can reject a conditional reset command, but not a forced reset command.

10.2.3.6.6 Cloudified NF Registration Management Service This service is defined in O-RAN.WG10.O1-Interface.0-R003-v12.00 [2], Section 6.10.

The Provider of this service is a cloudified Network Function (NF) instance. The Service Consumer is a management function responsible for the NF configuration and activation. Usually, the Consumer belongs to the SMO. Registration is a process of establishment of relation between the Service Consumer and the Service Provider.

Initialization of the Registration Management Service Provider includes setting up the NETCONF server and creation of the corresponding Management Object Instances (MOIs) representing the Service Provider as an NF. The address of the Consumer (IP or FQDN) should be made available to the Provider; for example, it can be pre-provisioned.

The registration process includes that the MnS Provider repeatedly sends a request for registration via O1 to the MnS Consumer, until complete establishment of the NETCONF session. The establishment of the session includes registration of the MnS Provider at the MnS Consumer. Further management operations performed by the MnS Consumer bring the MnS Provider to operational state.

10.2.4 Use Cases Supported in the O-RAN Operations and Maintenance Architecture

The document O-RAN.WG10.OAM-Architecture-v11.00 "O-RAN Operations and Maintenance Architecture" [3] provides Stage 2 description of several end-to-end OAM Use Cases, including objectives, involved entities, step-by-step process descriptions, and sequence diagrams. The addressed use cases are:

- **O-RAN Provisioning Use Case**
  This use case assumes specific deployment model with Virtualized Network Function (VNF) implementation of centralized functions: Near-RT RIC, O-CU-CP, O-CU-UP. It should be noted that the O-RU is always implemented as Physical Network Function (PNF), while the O-DU can be realized as a PNF or VNF. The involved entities are SMO, NFO, OAM Functions, Non-RT RIC, O-Cloud DMS, PNFs, and VNFs.
  The goal of the use case is to create an O-RAN network. To achieve this goal, the SMO may trigger instantiation of VNFs, addition of PNFs to inventory and setting interconnection between the NFs. The SMO activates Provisioning Service and performs configuration or reconfiguration of the Network Functions. When necessary, SMO may execute lifecycle management operations to add/remove/modify the involved NFs and perform software management actions.
- **O-RAN Measurement Data Collection Use Case**
  In this scenario, the Non-RT RIC performs control and/or optimization of the network, based on collection of measurement data for analysis and further actions. For this purpose, the Non-RT RIC requests the SMO OAM Functions to

arrange data collection. It is assumed that the SMO supports the corresponding MnS components per 3GPP TS 28.533 [25]:
- Necessary management operations and notifications ("MnS component Type A")
- The performance/fault information of the managed entity ("MnS component Type C")

It is also assumed that above MnS components are supported by the managed entity, playing the role of MnS Producer.

- After reception of the request from the Non-RT RIC, the SMO OAM Function initiates creation of a Measurement job at the MnS Producer (3GPP TS 28.550 [26]) and sets a subscription for the data to be generated and reported by the entity. When the SMO receives the data from the MnS Producer, this data can be accessed also by the Non-RT RIC.
- **O-RAN Configuration use case**
  The objective of the Configuration use case is to configure an NF (MnS Provider) in the way that the NF generates a Configuration Management (CM) notification toward the MnS Consumer every time when the configuration change. The MnS Consumer can also read the NF configuration. According to the document O-RAN.WG10.O1-Interface.0-v12.00 [2], the operations are based on NETCONF. The configuration changes, that trigger the notifications, can be result of operations dealing with Managed Object Instances (MOIs), such as Create Managed Object Instance, Delete Managed Object Instance, Modify Managed Object Instance. For receiving the CM notifications, the MnS Consumer subscribes to the MnS Provider using the Subscription Control operation.

10.2.5 O-RAN Management Information Models

The document "O-RAN Information Model and Data Models Specification" [7] defines mapping of O-RAN classes/components to 3GPP specifications. For example, the class ManagedElement used in O-RAN specifications, is defined in 3GPP TS 28.622 [6]. The O-RAN information model includes "by reference" definitions provided in 3GPP TS 28.541 [5]. From this point of view, O-RAN management information model can be considered an extension of the 3GPP management information model.

The following O-RAN specifications provide details of information models for specific O-RAN components:

- For O-CU-CP and O-CU-UP, see O-RAN.WG5.O-CU-O1.0-R003-v07.00 [8]
- For O-DU, see O-RAN.WG5.O-DU-O1.0-R003-v09.00 [9]

For example, the O-DU management specification [9] describes the following information model aspects, with mapping to corresponding 3GPP specifications where relevant:

- PNF Startup and Registration Management (with NETCONF realization details)
- Heartbeat Management Services
- PNF Software Management
- Performance Assurance Management
- Fault Supervision Management
- File Management
- Synchronization Aspects
- Provisioning Management
- Fronthaul Delay Management Support
- Trace Management

10.3 Open Fronthaul Management

10.3.1 M-Plane Architecture

Management Plane (M-Plane) of the Open Fronthaul interface is used for management of the O-RU. The M-Plane is based on NETCONF/YANG and supports Configuration Management (CM), Performance Management (PM), Fault Management (FM), and software management. As a special feature, there is support for O-RU installation.

Figure 10.2 shows two M-Plane architectural models. In the Hierarchical model, O-RU management goes through the O-DU. The management information related to the O-RU, goes between the SMO and the O-DU; the O-DU performs mapping between this information and NETCONF/YANG operations applied to the O-RU. In the Hybrid model, the SMO contains a NETCONF client, which communicates directly with the NETCONF server in the O-RU. This communication may go transparently via the O-DU device; in such case, the O-DU device is playing the role of IP router. In many cases, common name "O-RU Controller" is used for NETCONF clients in the O-DU and SMO.

Figure 10.3 shows the M-Plane protocol stack. The functionality of the "Application" layer includes Configuration Management (CM) and asynchronous notifications sent by the O-RU NETCONF Server. As an option, the M-Plane interface may support additional protocol stack configuration with JSON/REST asynchronous notifications on top of HTTP/TLS.

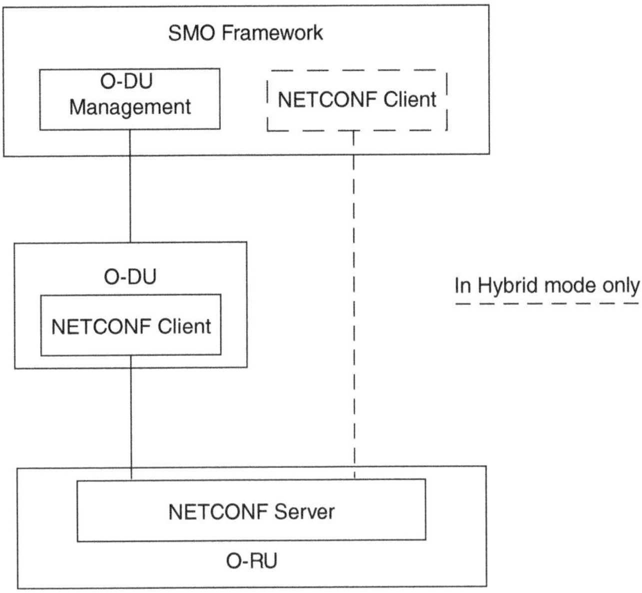

Figure 10.2 M-Plane Architecture. (*Source:* Vladimir Yanover.)

Figure 10.3 M-plane protocol stack. (*Source:* Vladimir Yanover.)

Application
NETCONF
SSH or TLS
TCP/IP
Ethernet/VLAN
Physical layer

10.3.2 M-Plane Functional Description

M-Plane specification O-RAN.WG4.MP.0-R003-v14.00 [1] is using the term "O-RU Controllers" for the elements managing the O-RU. Depending on the deployment scheme, it can be O-DU, SMO, or another management function.

The M-Plane supports the following functions:

- **"Start-up" installation**
  To establish IP connectivity to the O-RU Controller after power-up, the O-RU needs initial configuration, e.g. the IP address of the O-RU Controller. This address can be pre-configured in the O-RU itself, or the O-RU can receive it via DHCP. Next step is to establish secure connection from the O-RU NETCONF

Server to the NETCONF Client using the NETCONF "call home" procedure per IETF RFC 8071 [27]. In the Hybrid mode, the O-RU also receives the IP address of the SMO in same way, which also includes capability exchange.

- **Software management**
  This functionality supports download and activation of new software, initiated by the O-RU Controller.
- **Configuration management**
  The M-Plane is using NETCONF <get-config> and <edit-config> RPCs (IETF RFC 6241 [11]) for retrieval and update of the O-RU configuration parameters.
- **Performance management**
  Performance Management (PM) performs collection of network data, such as performance measurements. For data collection, M-Plane is using the <get> operation (IETF RFC 6241 [11]). Another possibility is data collection in a file, which is subsequently uploaded to the O-RU Controller.
- **Fault management**
  For delivery of the alarm information, alarm notifications are sent to the O-RU Controller
- **File management**
  As a part of the File Management, the O-RU Controller can trigger transfer of the files (e.g. with performance data or alarm list) from the O-RU to the O-RU Controller.

References

1 O-RAN.WG4.MP.0-R003-v14.00. O-RAN Fronthaul Management Plane Specification.
2 O-RAN.WG10.O1-Interface.0-R003-v12.00. O-RAN Operations and Maintenance Interface Specification.
3 O-RAN.WG10.OAM-Architecture-v11.00. O-RAN Operations and Maintenance Architecture.
4 3GPP TS 28.530. Management and orchestration; Concepts, use cases and requirements.
5 3GPP TS 28.541. 5G Network Resource Model (NRM); Stage 2 and stage 3.
6 3GPP TS 28.622. Generic Network Resource Model (NRM). Integration Reference Point (IRP); Information Service (IS).
7 O-RAN.WG10.Information Model and Data Models.0-R003-v07.00. O-RAN Information Model and Data Models Specification.
8 O-RAN.WG5.O-CU-O1.0-R003-v07.00. Interface specification for O-CU-UP and O-CU-CP.
9 O-RAN.WG5.O-DU-O1.0-R003-v09.00. O1 Interface specification for O-DU

10 O-RAN.WG1.OAD-R003-v11.00. O-RAN Architecture Description.
11 IETF RFC 6241. Network Configuration Protocol (NETCONF).
12 IETF RFC 7950. "The YANG 1.1 Data Modeling Language".
13 3GPP TS 28.531. Management and orchestration; Provisioning.
14 3GPP TS 28.532. Management and orchestration; Generic management services.
15 OPNFV. (Anuket) Copper Project, https://artifacts.opnfv.org/ves/review/design/index.html (accessed 23 May 2024)
16 3GPP TS 28.545. Fault Supervision (FS).
17 3GPP TS 28.623. Generic Network Resource Model (NRM) Integration Reference Point (IRP); Solution Set (SS) definitions.
18 3GPP TS 28.552. 5G performance measurements.
19 3GPP TS 32.421. Subscriber and equipment trace; trace concepts and requirements.
20 3GPP TS 32.342. File Transfer (FT) Integration Reference Point (IRP); Information Service (IS).
21 3GPP TS 32.346. File Transfer (FT) Integration Reference Point (IRP): Solution Set (SS) definitions.
22 3GPP TS 28.314. Management and orchestration; Plug and Connect; Concepts and requirements.
23 3GPP TS 28.315. Management and orchestration; Plug and connect; procedure flows.
24 3GPP TS 28.316. Management and orchestration; Plug and Connect; Data formats.
25 3GPP TS 28.533. Management and orchestration; Architecture framework.
26 3GPP TS 28.550. Performance assurance.
27 IETF RFC 8071. NETCONF Call Home and RESTCONF Call Home.

11

AI/ML Support in O-RAN Architecture

CONTENTS

Acronyms and Abbreviations

AI	Artificial Intelligence
AI/ML	Artificial Intelligence/Machine Learning
CA	Carrier Aggregation
CN	Core Network
C-SON	Centralized SON
CU	RAN Central Unit
DC	Dual Connectivity
DL	Downlink
DRB	Data Radio Bearer
D-SON	Distributed SON
DU	RAN Distributed Unit
FCAPS	Fault, Configuration, Accounting, Performance, Security management functions
gNB	5G Base Station defined in 3GPP
H-SON	Hybrid SON

Open RAN: Technology and Ecosystem, First Edition. Vladimir Yanover.

LCM	Life Cycle Management
MDP	Markov Decision Process
ML	Machine Learning
ms	millisecond
Near-RT RIC	Near-Real-Time Radio Intelligent Controller
NF	Network Function
Non-RT RIC	Non-Real-Time Radio Intelligent Controller
O1	O-RAN management interface (OAM, FCAPS)
O-CU	O-RAN defined Central Unit
O-CU-CP	O-RAN defined Central Unit – Control Plane
O-CU-UP	O-RAN defined Central Unit – User Plane
O-DU	O-RAN defined Distributed Unit
OFH	Open Fronthaul interface
O-gNB	O-RAN conformant 5G Base Station, which, in particular, supports E2 interface
O-RU	O-RAN defined Radio Unit
PM	Performance Management
RAN	Radio Access Network
RT	Real Time
RU	Radio Unit
SMO	Service Management and Orchestration framework
SON	Self-Organized Network
UE	User Equipment
UL	Uplink

11.1 Introduction

This chapter describes O-RAN architecture aspects related to use of Artificial Intelligence/Machine Learning (AI/ML) in RAN automation/optimization, including placement of associated components and the structure of control loops. The chapter contains classification of AI/ML algorithms and description of the lifecycle of AI/ML models, including ML model training, deployment, and inference. For selected AI/ML architecture options, possible locations for AI/ML model training and inference ("deployment scenarios") are discussed, with mapping of AI/ML components to the O-RAN control loops.

The chapter also describes O-RAN approach to Federated Learning implementation, including collaborative training of an AI/ML model by multiple AI/ML entities spread over the O-RAN system.

A novel RAN architecture [1] is one of most important O-RAN technology innovations. The O-RAN architecture includes several aspects, which leverage RAN control and optimization capabilities. The architecture contains centralized non-real-time and near-real-time controllers, Non-Real-Time Radio Intelligent Controller (Non-RT RIC), and Near-Real-Time Radio Intelligent Controller (Near-RT RIC). These controllers can be used for centralized RAN control and optimization. They also can be used as platforms hosting 3rd-party applications that implement control and optimization algorithms; see [2, 3].

The Service Management and Orchestration framework (SMO) supports O-RAN management interface O1, which can be used also by Non-RT RIC. The O1 interface provides for full support of Fault, Configuration, Accounting, Performance, Security (FCAPS) management operations defined in 3GPP (see [4–9])

Coordination between the controllers is based on the A1 interface, which allows Non-RT RIC to set the policies for Near-RT RIC operations and deliver to the Near-RT RIC the Enrichment Information ([10–12]). The Near-RT RIC can use the E2 interface to receive from subordinated RAN components ("E2 Nodes") network performance indicators (E2 REPORT Service) and real-time notifications of internal events (E2 INSERT Service). Based on analysis of the received information, the Near-RT RIC can issue real-time commands changing the behavior of RAN and/or initiating specific RAN procedures (E2 CONTROL Service) or set real-time policies to be executed by RAN components (E2 POLICY Service). More information can be found in Chapter 7 and specifications [13–17].

Combined, above capabilities of the Non-RT RIC and Near-RT RIC allow a wide range of RAN control and optimization procedures, enabling efficient support of strong use cases; see overview of the use cases in Chapter 9 and specifications [12, 18, 19].

11.2 Control Loops

O-RAN specification O-RAN.WG2.AIML-v01.03 [20] defines three types of control loops, which differ in their time scale (Figure 11.1).

Loop 1 type, controlled by the O-DU, operates in the TTI time scale: below 10 ms. Example of Loop 1 operation: Layer 2 scheduling. Loop 2 operates in the Near-RT RIC in the scale of 10–1000 ms. Example of controlled processes: mobility procedures, such as handover, Dual Connectivity and Carrier Aggregation configuration changes. Loop 3 is controlled by the Non-RT RIC at the time scale

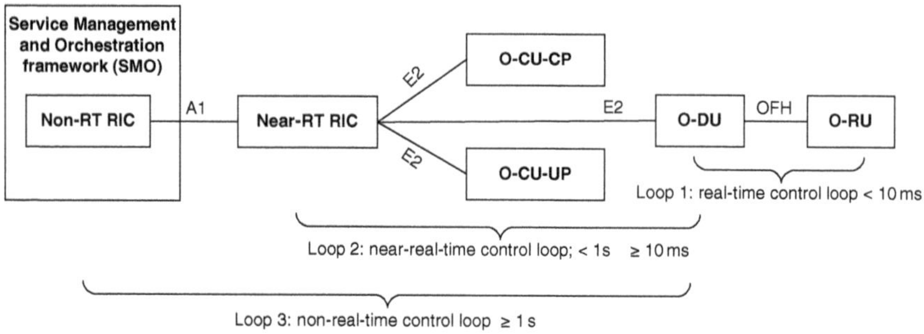

Figure 11.1 Control loops in O-RAN. (*Source:* Vladimir Yanover.)

greater than 1000 ms. Examples of operations: lifecycle management of A1 Policy, O-Cloud orchestration.

Some solutions include nested "slow" and "fast" loops, where the "slow" loop may control the configuration parameters of the "fast" loop.

11.3 AI/ML Model Lifecycle

Figure 11.2 shows AI/ML model lifecycle with the following state transition actions:

- Creation of the model
  The model is created, but not trained yet.
- Training data preparation
  Preparation of data set needed for model training
- Model training and validation
  Model training, based on the training data set. After training finished, model validation may be performed by applying the model to different data set ("validation data set")
- Model selection and deployment
  Model is selected for inference and deployed at the inference engine, but not activated yet
- Model activation
  Model is activated for use in inference
- Model deactivation
  Model is deactivated
- Training data set update
  When there is a need in retraining the model, updated training data can be collected; the model returns for retraining
- Model deletion
  Model is deleted

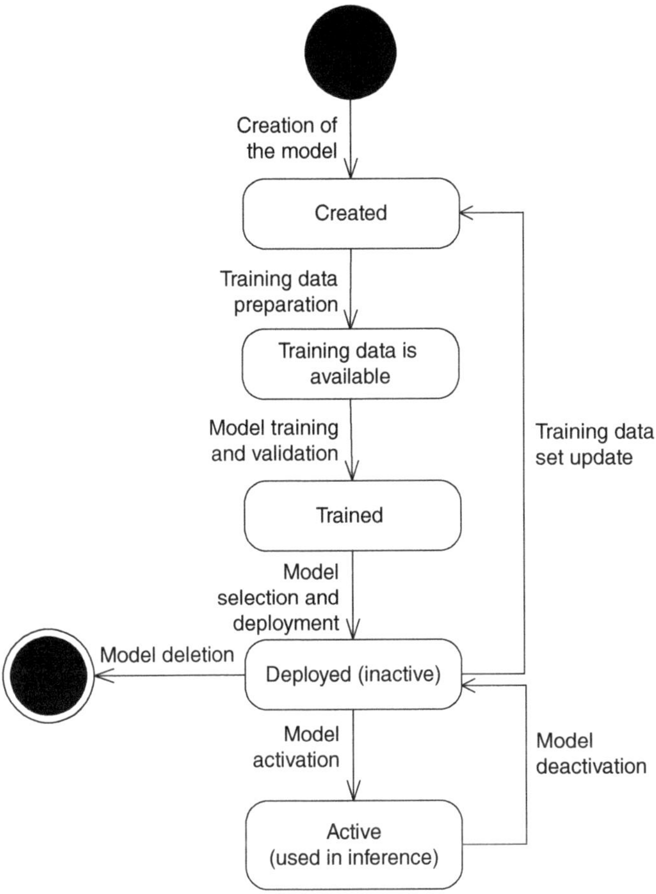

Figure 11.2 AI/ML model lifecycle. (*Source:* Vladimir Yanover.)

11.4 AI/ML Model Deployment Scenarios

Figure 11.3 shows two examples of AI/ML deployment scenarios defined in the O-RAN specification O-RAN.WG2.AIML-v01.03 [20]. The scenarios differ by location of AI/ML-related functionalities and use of specific interfaces for transfer of AI/ML-related information.

The labels used in the figure:

- Functionalities
  - "Data prep": preparation of data input
  - "Training": training of the AI/ML model
  - "Inference": inference executed, based on the trained model

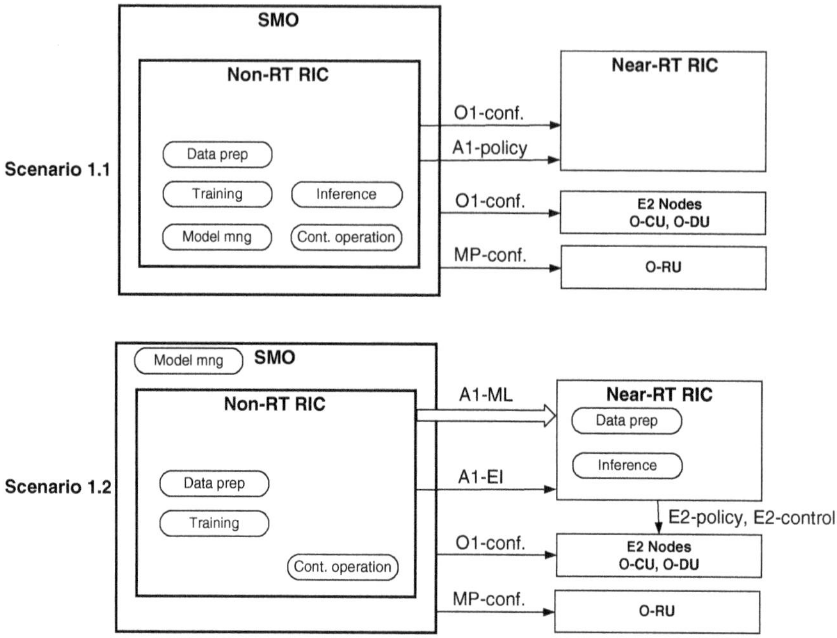

Figure 11.3 AI/ML model deployment scenarios. (*Source:* Vladimir Yanover.)

  - "Model mng": AI/ML model management
  - "Cont. operation": continuous improvement of the AI/ML model; includes verification /monitoring/analysis/recommendation/optimization.
- "**A1-ML**": delivery of the trained AI/ML model via the A1 interface
- Configuration/policy interfaces
  - "O1-conf": Configuration Management (CM)
  - "MP-conf": Configuration Management via the M-Plane of the OFH interface; the figure shows Hierarchical deployment model (Chapter 10)
  - "A1-policy": Management of A1 Policies via the A1 interface (see Chapter 6)
  - "A1-EI": delivery of Enhancement Information via the A1 interface
  - "E2-policy": use of E2 POLICY service (see Chapter 7)
  - "E2-control": use of E2 CONTROL service (see Chapter 7)

In Scenario 1.1, the data preparation, ML model training, continuous operations, and inference are performed in the Non-RT RIC. In this location, the Inference functionality is naturally using control loop 3 with time scale >1 seconds (Figure 11.1). The Non-RT RIC executes optimization actions via the O1 interface, toward O-RAN network functions: Near-RT RIC/ O-CU/ O-CU-CP /O-CU-UP/ DU. Another type of optimization actions is setting/modification of the A1 Policies at the Near-RT RIC via the A1 interface.

In Scenario 1.2, data preparation is performed in both Non-RT RIC and Near-RT RIC. Training of the ML model and continuous operations are performed by the Non-RT RIC. The trained model is delivered to the Near-RT RIC via the A1-ML interface. Location of Inference in the Near-RT RIC naturally implies use of control loop 2 with time scale between 100 ms and 1 second (Figure 11.1). The Near-RT RIC executes optimization actions toward E2 Nodes via E2 interface, using E2 POLICY and E2 CONTROL services. Enrichment Information can be delivered to the Near-RT RIC via the A1-EI interface.

O-RAN specification O-RAN.WG2.AIML-v01.03 [20] provides some examples of AI/ML algorithms, which can be implemented in the O-RAN architecture.

Supervised Learning examples, where the training data set is labeled:

- **Regression**: Linear Regression, Logistic Regression
- **Instance-based Algorithms**: k-Nearest Neighbor (KNN)
- **Decision Tree Algorithms**: CART
- **Support Vector Machines**: SVM
- **Bayesian Algorithms**: Naive Bayes
- **Ensemble Algorithms**: Extreme Gradient Boosting, Bagging: Random Forest

Examples of Unsupervised Learning with the training data set not labeled:

- K-means clustering
- Principal Component Analysis (PCA).

11.5 Reinforcement Learning

Reinforcement Learning (RL) is a kind of AI/ML algorithms, based on "trial – error" interaction in changing environment. O-RAN specification O-RAN.WG2.AIML-v01.03 [20] points to some examples:

- Q-learning
- Multi-armed bandit learning
- Deep Q Network
- State-Action-Reward-State-Action (SARSA)
- Temporal Difference learning
- Actor–critic reinforcement learning
- Deep deterministic policy gradient
- Trust region policy optimization
- Dyna-Q
- Monte Carlo tree search

Figure 11.4 shows possible deployment scenario for the case of RL. In this scenario, initial model training is done in the Non-RT RIC. After the trained model is delivered

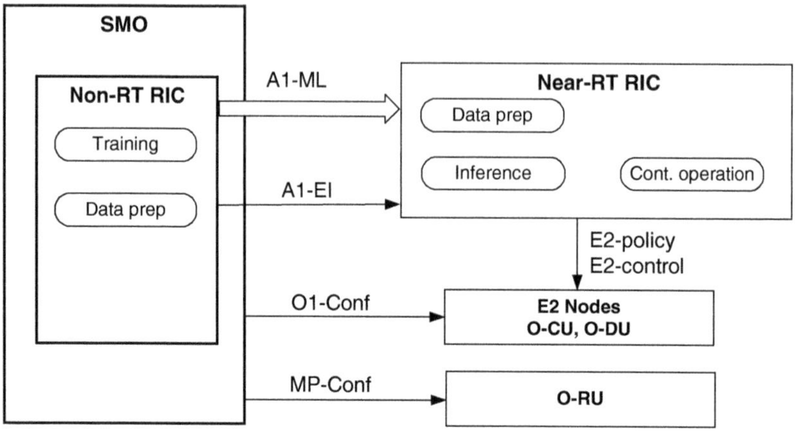

Figure 11.4 RL model deployment scenario. (*Source:* Vladimir Yanover.)

to Near-RT RIC, the RL engine is using the trained model for inference, at the same time performing model optimization and/or re-training ("Continuous operation").

11.6 Federated Learning

Federated learning is a kind of distributed learning, where multiple AI/ML learning engines collaborate in training of common AI/ML model. Every learning engine performs training on the available local portion of training set, which results in a locally trained model. The locally trained models are delivered to the central training host for integration.

In O-RAN architecture, Non-RT RIC can play role of the central training host with several Near-RT RICs performing local training and inference. Figure 11.5 shows example of deployment scenario, which can be used for this purpose. In this scenario, data preparation is performed at the Near-RT RIC. The Non-RT RIC creates initial ML model, which optionally can be trained, e.g. on historical data. A copy of the model is delivered for local training to every Near-RT RIC via the interface A1-ML-Dn. Local training is performed in every Near-RT RIC separately, based on local training data set. To create the local training data set, the Near-RT RIC performs network data collection via E2 interface using REPORT and INSERT services (not shown in the figure). Locally trained model is delivered back to the Non-RT RIC via the A1-ML-Up interface. Next step is integration of multiple locally trained models in the Non-RT RIC. A copy of the integrated model is delivered to every Near-RT RIC, where it is used for inference. The Near-RT RIC executes optimization actions toward RAN components (E2 Nodes) via the

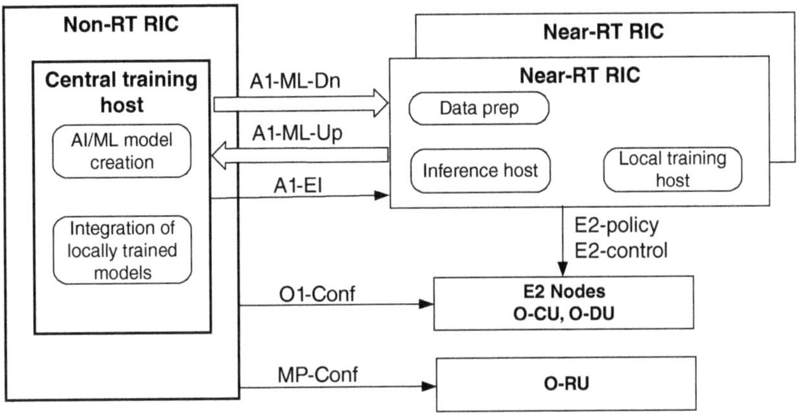

Figure 11.5 Federated learning deployment scenario. (*Source:* Vladimir Yanover.)

E2 interface, using E2 POLICY and E2 CONTROL services. Enrichment Information can be delivered to the Near-RT RIC via the A1-EI interface.

Figure 11.6 shows the process flow. The process begins with AI/ML model creation and possible pre-training in the central training host in the Non-RT RIC. Then the central training host distributes the model to local training hosts, in

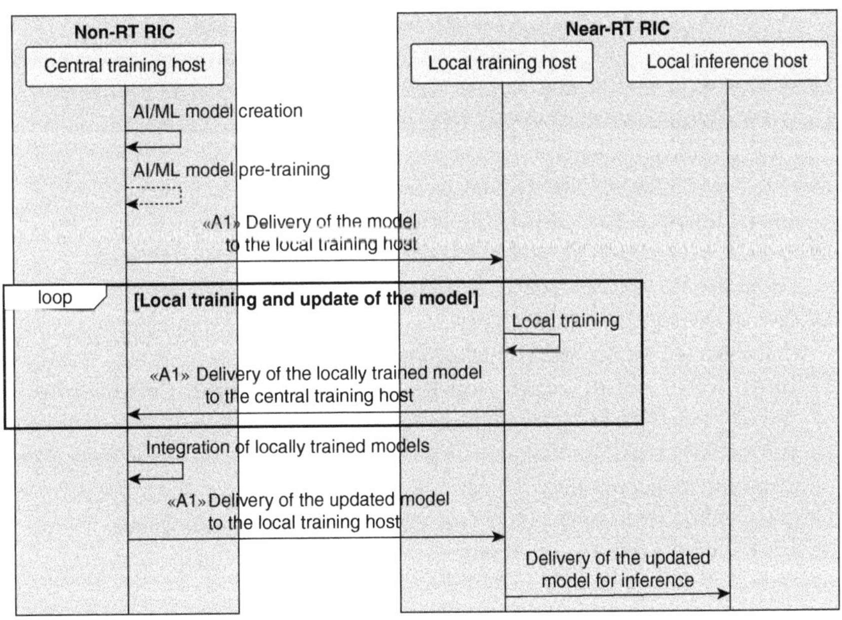

Figure 11.6 Federated learning: process flow. (*Source:* Vladimir Yanover.)

which the model is additionally trained on the local data set. Then the result is sent back to the central training host for integration. When all locally trained models are integrated, the updated model is distributed to local inference hosts in Near-RT RICs. The integrated model is used for inference in the local inference hosts.

References

1 O-RAN.WG1.OAD-R003-v11.00. O-RAN Architecture Description.
2 O-RAN.WG2.Non-RT-RIC-ARCH-R003-v05.00. Non-RT RIC Architecture.
3 O-RAN.WG3.RICARCH-R003-v05.00. Near-RT RIC Architecture.
4 3GPP TS 28.531. Management and orchestration; provisioning.
5 3GPP TS 28.532. Management and orchestration; generic management services.
6 3GPP TS 28.541. 5G Network Resource Model (NRM); Stage 2 and stage 3.
7 O-RAN.WG10.Information Model and Data Models.0-R003-v07.00. O-RAN Information Model and Data Models Specification.
8 O-RAN.WG10.O1-Interface.0-R003-v12.00. O-RAN Operations and Maintenance Interface Specification.
9 O-RAN.WG10.OAM-Architecture-R003-v11.00. O-RAN Operations and Maintenance Architecture.
10 O-RAN.WG2.A1AP-v04.01. A1 interface: application protocol.
11 O-RAN.WG2.A1GAP-R003-v03.02. A1 interface: General Aspects and Principles.
12 O-RAN.WG2.A1UCR-R003-v01.02. A1 interface: Use Cases and Requirements.
13 O-RAN.WG3.E2AP-R003-v05.00. Near-Real-time RAN Intelligent Controller, Use Cases and Requirements.
14 O-RAN.WG3.E2GAP-R003-v05.00. Near-Real-time RAN Intelligent Controller and E2 Interface. E2 General Aspects and Principles (E2GAP).
15 O-RAN.WG3.E2SM-CCC-R003-v03.00. Near-Real-time RAN Intelligent Controller. E2 Service Model (E2SM), Cell Configuration and Control.
16 O-RAN.WG3.E2SM-KPM-R003-v04.00. Near-Real-time RAN Intelligent Controller. E2 Service Model (E2SM) KPM.
17 ORAN.WG3.E2SM-RC-v05.00. Near-Real-time RAN Intelligent Controller. E2 Service Model (E2SM), RAN Control.
18 O-RAN.WG2.Use-Case-Requirements-v09.00. Non-RT RIC & A1 Interface: Use Cases and Requirements.
19 O-RAN.WG3.UCR-R003-v05.00. O-RAN Use Cases and Requirements.
20 O-RAN.WG2.AIML-v01.03. AI/ML workflow description and requirements.

Index

a

Open RAN: Technology and Ecosystem, First Edition. Vladimir Yanover.

b

c

f

g

p

S

www.ingramcontent.com/pod-product-compliance
Ingram Content Group UK Ltd.
Pitfield, Milton Keynes, MK11 3LW, UK
UKHW022038190125
453848UK00007B/21

9 781394 241149